THE SEA

John Crompton

THE SEA

WITH 24 DRAWINGS BY DENYS OVENDEN

NLB

Nick Lyons Books

Printed in the United States of America

10 9 8 7 6 5 4 3 2 1

Published by arrangement with Doubleday & Co.

Library of Congress Cataloging-in-Publication Data

Crompton, John, 1893–
  The sea.

  Includes index.
  1. Marine biology. I. Title.
QH91.C76 1988   591.92   88–538
ISBN 0-941130-83-5

CONTENTS

INTRODUCTION

Having read Crompton on the Spider, the Snake, the Bee, the Ant and the Hunting Wasp, I was puzzled—and a bit worried—to learn he'd essayed a book on the Sea. The charm of the one-critter books, it seemed to me, lay in the intensity of their focus, the odd, often chilling minutiae Crompton could find in the hidden corners of his subjects' lives. Would the Sea, in its vastness, dilute the macabre Cromptonian wit? How could the author's delightful, low-keyed introspection—peppered as it always is by laconic throwaway bits of his own weird experiences in China, Africa and rural England—survive so wet and churning a voyage? Would he sink or swim?

I needn't have fretted. Crompton on the Sea is Crompton to the hundredth power. For the Sea, as the good book stateth, is full of wonders. And they are just the sort of wonders Crompton relishes most: strange, fierce, clever creatures evolving and perishing and starting again, working their wily strategies of survival in an environment so bizarre as to make George Lucas's Industrial Light & Magic Company seem like a second-rate toy shop. Consider, for instance, the following.

"A certain species, of the angler-fish tribe, judging by all the specimens secured, possessed only females. No males were ever found, even when such ought to have been hanging around. It was rather a mystery. Where was the husband of this female who regularly produced her fully fertilised eggs? The truth, when it came to light, was indeed surprising. The male, when very young and very small, bites the skin of the female, and holds on. He holds on so tenaciously that in time he becomes embedded in her flesh and her skin grows over the aperture and imprisons him. He is now a part of her and gradually becomes even more so. Her veins merge into his. His heart and digestive system decay away (he has no need for them for her blood flows through him and nourishes him). In his almost empty body his testes enlarge and develop like some cancerous growth until his whole inside is crammed. To all intents and purposes he is now merely a bag of sperm, and the female can be said to contain both ovaries and testicles. For she can draw on his sperm as she desires whenever she lays her eggs. Never in nature has female possessiveness gone further."

Or this: "The barnacle starts life with one of the largest eyes possessed by any living creature, and finishes up with no eyes at all." Either of these quotes could serve as an epigraph to some cryptic Symbolist novel, but just when Crompton begins to seem a bit too metaphorical, he throws in something like this from his fund of deep sea lore: "As an example of the adequacy of the cement they use . . . barnacles have been found on ships' propellors after a long voyage. Most of us have had from time to time uncomfortable sea trips, but I can imagine no more uncomfortable a journey than when travelling on the blade of a propellor!"

Pure Gary Larson, but written long before the daft draughtsman of *The Far Side* graduated from Crayola-chewing. In fact there are a lot of oddball, Larsonesque touches in Crompton—the hermit crab that carries a stinging sea anemone on its back as a defense against undersea muggers, the young of the European oyster taking "their first swimming practices" inside their mother's fluid-filled shell, and the aftermath, once they have learned. "When they depart, no goodbyes are said, and there is no lingering; the mother slightly opens her shell and spits them out like a charge of pellets from a shotgun." Or, slightly more grotesque, the hapless Indian (I imagine him as one of those sanctimonious bank clerks in Bombay, sneering as he shortchanges your rupees) who while "enjoying a quiet bathe in the sea . . . was cut clean in half by one slash from a saw fish."

Yes, there are monsters in Crompton's Sea—and it's refreshing to find them there in our wimpy age. For all their technical brilliance, the nature films on television, from which most modern folks glean what little they know of the natural world, would have us believe that all creatures are not only wise and wonderful, but harmless if not indeed downright kindly as well. Crompton, product of a tougher time, delights in a Nature red in tooth and claw. His sharks eat Newfoundland dogs for canapés. His whales don't sing; they smash boats and chomp men. His giant squids are "as big as tennis-courts" or else bear uneasy resemblance to "one of Hitler's V2 rockets that used to be exhibited in Trafalgar Square towards the end of the last war."

He waxes positively gleeful over the shocking power of the electric ray ("the effect on a man who trod on one would be spectacular") and delights in the creature's existence: "Nature, in fact, rather surprisingly, has shown herself here (and in other electric fishes) to be a skilled and inventive electrician. I say surprisingly because with most other animals she has given no hint that she knew anything much about the subject, and in the inanimate world her best known efforts—thunderstorms—are rather crude affairs."

Yet he can chide Nature, too, for her whimsical unfairness. "I have already commented on the injustice of giving replaceable teeth to fishes and reptiles but not to us. Now we find that she has given replaceable limbs to mere lobsters and other crustaceans but again not to us. Yet a lobster has ten legs against our two. And there is worse to come. Nowadays artificial teeth and limbs can to some extent replace missing members, but there is no substitute for an eye. Cut out a man's eyes and he will never see again, but cut off a lobster's eyes and nature will grow new ones for him, complete with the thirteen thousand lenses and thirteen thousand nerve rods, and as good in every respect as the originals." (Tell *that* to your optometrist.)

There is nothing wishy-washy in Crompton's Sea, none of that limp-wristed, even-handed "relativism" on which the closing of the American mind has recently been blamed. He calls things as he sees—and feels—them, in strong, clear, vibrant English untainted by scientific jargon. "The whelk is the rat of the sea," he says. "Nothing alive seems to come amiss to it, and it often works havoc on an oyster bed by gnawing into the shells of the oysters and sucking out the contents . . . . A nasty creature, the whelk, without even

the saving grace of being eatable. I am glad to say that not everyone
agrees with me here and immense numbers of whelks are sold at
seaside stalls and eaten by holiday-makers so that large numbers are
caught and taken away from the scene of their nefarious activities,
but I still maintain that the whelk is uneatable—I would as soon
chew a rubber tyre."

No, you don't go to Crompton for the latest in oceanographic
news—no more than you'd go to Ray Bergman or John Waller Hills
for the latest in trouting techniques or fly theory. You'll find no
mention here of the role plate tectonics played in the formation of
the seas. Writing when he did, some thirty years ago, Crompton had
no knowledge of it. He grossly underestimates even the age of the
earth (by two billion years, no less) and is vague if not indeed mis-
leading on the subject of mass extinctions. His version of the dinosaurs' disappearance owes more to Walt Disney's *Fantasia* than to
current theories of collision between the earth and the so-called
planet Lucifer. You'll learn more about the current state of thinking
regarding the sea by watching Jacques Cousteau, and certainly
Rachel Carson's fine sea trilogy, written about the same time as
Crompton's book, is far more comprehensive.

But you don't read Crompton for that sort of thing. You read him
for his inimitable voice—those marvelous locutions, antique in the
finest sense. For the gruff wit, the crabbed, offbeat insights that snap
your head around two paragraphs later, the rambling enthusiasms of
a man who loves (and loves to convey to his readers) the rich, wacky,
infinite variety of life on this marvelous planet. His books are Besti-
aries for our time, full of mystery, awe, wonder and a delight in even
the creepiest of the crawlies the Earth can produce. That's nature
writing at its most enjoyable, for in it lies revealed not just the nature
of the planet, but of the writer himself.

—ROBERT F. JONES
January, 1988

FOREWORD

THE most remarkable thing about the sea is that it is there at all. None of our fellow-planets possess a sea. So, for this reason alone, it is difficult to think that they can support much life. For our own sea is the source of all our water, and living things are composed chiefly of water. Furthermore, life originated in the sea, developed in the sea, and spread to land from the sea.

I touch on these aspects of the sea in the first few chapters; the remainder of this book deals with representatives of the teeming life that is in the sea today. The subjects selected are mostly well known and their life histories have been studied.

And that, I think, is enough by way of an explanation of the scope and limits of this book.

There remains to thank the many authorities who have given me information and cleared up doubtful points. I also wish to thank Mr. R. S. R. Fitter who was kind enough to go over the first draft of the manuscript and give invaluable help and criticism.

THE SEA

CHAPTER 1

LIFE AND THE WATER

In the first chapter of Genesis come these words: "Let the waters bring forth abundantly the moving creature that hath life, and fowl that may fly above the earth . . ."

Many scribes holding differing views on religion and other matters undoubtedly contributed to the book of Genesis. Did one of these hold the view that all life came from the sea? If so, he was far in advance of modern thought up to a hundred or so years ago. But we will go into that later; first let us speculate about the world before water or life came.

We do not know how the world was made. There are theories in number. Some think that it originated from the collision of a comet with the sun; some from the collision of another star with the sun, the planets being the bits and pieces that were hurled into space; some that a star once passed close by the sun, raising waves on its nearest surface which broke off into separate bodies, some of which departed with the interloper while others stayed with the sun, circling round her. Some believe that the sun was once a mass of whirling gas exactly the size of our present planetary system, which condensed and in condensing threw off whorls which themselves condensed into planets, sub-planets, and dust. Others maintain that the planets (being of different material) did not come from the sun at all, but from a companion star that exploded, a supernova which left clouds of incandescent gas near the sun before it whirled away into outer space. And so on. For, as a scientist said recently, "Every few years a new theory of the origin of the earth appears."

In short, we do not know.

We are on surer ground, however, in our computations of the age of the world. From astronomical and geological considerations and with the help of uranium it is now commonly accepted that the world came into being about 2,500 million years ago. This must not, of course, be taken as a pinpoint in time like the dates in history books. It may have been less, it may have been considerably more; the remarkable thing is that it has been possible to get as near as we have.

In previous computations theology had its influence. Man has always regarded God as his own special possession and has been jealous of any others who might seem to have been equally favoured. Therefore, he could not think of a world in which he lived that had not been specially created for him. He took it for granted —yes, and not many years ago—that he was the principal figure in the creation. He allowed the Creator a certain measure of time to create and get the world ready for him, but he only allowed Him a few days. (Whoever or whatever created the world *did*, in a sense, get it ready for man, but it took longer than a few days.)

And so we find, in 1654, an Irish Archbishop announcing that, after intensive study of the scriptures, he had found that the world was created on twenty-sixth October, 4004 B.C. at nine o'clock in the morning. And for a hundred years this was taken as official. Later, other dates were given for the birth of the world, but most of them brought the date even nearer the present day than those calculated before.

The opinion of the majority of American and European authorities is that the world was incandescent to start with, and that even if it started cold, various agents including radio activity and gravity soon "set it on fire." Assuming that these authorities are correct, the earth must have been a beautiful sight in that first period—a brilliant, star-like globe. But as it cooled it became a lake of dull-red molten rock belching smoke and sulphur fumes and illuminated by a firework display of meteors and volcanoes. Later it became even more hell-like in appearance as masses of black smoking slag rose and floated on the red surface. These were the beginnings of what we know now as the continents.

The ugly masses of slag began to settle on the solidifying heavier basalt rocks below. The continents were taking shape and fundamentally assuming the shapes they have today.

Meanwhile in the cooling process of the earth, water vapour in prodigious quantities was being manufactured and ejected. In the far-away cooler upper atmosphere it was allowed to condense as cloud, but if it ever fell as rain, as it must have done, it was wasted effort, for it fell on hot rock and immediately boiled and went back whence it had come. Rocks solidify at about 2000° F. Water boils at 212°. So the rocks had a long cooling process to go through before any water could fall on them without sizzling back into steam. A dense mass takes longer to cool than a smaller mass, and it must have been many ages after the solidifying of the rocks before their temperature fell below the boiling point of water. But at long last they *did* fall below that temperature and then—then the rain came. It was a cloudburst that continued without stopping for over a hundred years. It was rain such as the earth has never had since and never will again. The whole of the great oceans had for long been hanging over the earth in an unbroken dense black cloud. The surface of the world had never seen the sun for thousands of years; all it got was a thin filtering-through of very faint light in the daytime.

We started our picture of earth with a hell of molten rock and smoking slag. That scene has now changed for another almost as unpleasant: an unceasing thunderstorm. Were that storm to fall on us today it would destroy all land life, and it would sweep every particle of soil and goodness from the land into the sea. In those days, however, there was no soil and no goodness—only bare rock.

At a certain spot in the interior of Africa where I lived for some years there was a very pretty lake picturesquely surrounded by bluegum trees, where I and my companions used to bathe until the dry season dried the lake up, leaving only an ugly, deep, fissured pit. Then the dry season would break, black clouds would pile up, roars as from distant artillery would be heard, and soon the rain would come, piling up its many inches in the hour. In the morning the rain would be gone and the sun would be shining, but the empty crater would be a lake again, lapping against its banks.

That, in miniature, is what must have happened in the days of the first flood. After what must have seemed like perpetual night on earth the sun at last broke through the clouds and disclosed the glittering oceans.

The sea had been born.

After that comes never-ending time. A thousand thousand years follows a thousand thousand in endless procession. Continents shrink and swell. Mountains rise and are worn flat to the ground again. And the sea—

> *The Sea, unmated creature, tired and lone,*
> *Makes on the desolate sands eternal moan.*

Yes, after all those years, there was the sea, and there were the rocks and the sand. What else *could* there be?

But something mysterious happened; a thing so small that it probably could not have been seen even with the highest-powered microscope of today, got into the water. When we break matter down we can get all matter so small that it cannot be seen by a microscope, and earth and sea are made of bits like this. But this speck was different. It was matter like the other specks, but within it, like an imp in a bottle, was LIFE.

There is a right explanation of everything—if you can find it. There is an explanation of life and there is an explanation of how it came to earth, but nobody has found it. There have been guesses, and indeed assertions, but we are still no wiser than we were before these knowledgeable folk made their guesses and their assertions. Let us look at some of the explanations that have been given, not of what life *is*—that is too difficult—but of how it came to earth. There are those who ascribe it to God, and there is nothing that we know of against this theory. It may seem to our small and finite minds an intricate and prolonged way of creating, for instance, man, but to a Deity time is non-existent. At the moment, however, this is a material world and many men are beginning to reason in their own small way and to ask for proofs instead of statements. This may be a wrong attitude, but if it is, it is largely the fault of those theologists who in the past have jumped to hasty conclusions on inadequate data. Therefore, mankind at the present time is inclined to be dubious about their unproved findings.

Others believe that life came out of space in a meteor or some such vehicle. But meteors on striking our atmosphere become incandescent and usually disintegrate so that if life could survive those conditions then life on the surface of the sun is not an impossibility.

A more likely supposition is that life came to earth in a speck of dust; while some think that it came in the shape of spores from interstellar space, and do not rule out the possibility of their having been scattered by intelligent beings.

As is often pointed out, to believe that life came from outer space does not solve the problem. *Where* did it come from in outer space and how did it get *there?* It does not solve the problem certainly, but it would be satisfactory to know. It would be taking one step at least, instead of none.

The theory most commonly accepted is that life occurred spontaneously under conditions that can never happen again. We are not told, however, exactly what these conditions were and why they can never happen again. This theory, by the way, seems rather akin to the theories of pre-Pasteur days when it was generally believed that the smaller animals were spontaneously engendered from corruption, and when processes and recipes were given to—amongst other things—make fishes out of slime, spiders out of dust, and honey bees out of the decaying carcasses of bullocks. For instance Izaak Walton—"As pearls are made out of glutinous dewdrops, so eels are bred of a particular dew, falling in the months of May or June on the banks of some particular ponds apted by nature for that end."

"Apted by nature for that end" is vague and does not really commit Walton, nor does "conditions that could never possibly occur again" commit the more recent speculators. Walton, however, goes on, casting discretion aside: "Eels may be bred as some worms, and some kinds of bees and wasps, either of dew or out of the corruption of the earth."

Corruption of course does not apply to the days of which we write. To get corruption one must have life; the two go hand in hand and are mutually dependent on each other. Had a sterilised man been transported to that early era, and been able to exist without food, he would have found it almost impossible to get ill and very hard to die except by falling down a precipice and breaking his neck, or being drowned. There would be no blood-poisoning nor pneumonia, and when he died of sheer old age there would be no need for his friends (supposing he had any with him) to bury him. However hot the climate he would not decay. For decay is not a natural thing, but the remarkable invention of certain bacteria that first had to learn how, for their own ends, to accomplish this amazing chemical feat.

The theory we are examining also lays down that all life is the same age, that you and I, the dandelion on the lawn, the mite in the cheese, and the rest came from the life that appeared so mysteriously: that, in short, life never again came to earth (to survive) and never will.

The search goes on. In fact it has never been so frenzied as it is today. Professors and other learned men are concentrating their formidable knowledge in the hunt for the origin of life. So far, to judge by their lack of agreement, they have not yet sighted their objective. I quoted a scientist as saying that every few years a new theory of the origin of the earth appears and I think that the same thing could be said of the origin of life.

In the main (although other possibilities are not absolutely ruled out) the present-day investigators are working on the assumption (to put it ridiculously inadequately) that life came into being from molecular build-ups and changes, and that life is a thing that is bound to come about, given the right conditions. One thing must be said: never before has this problem been tackled by men possessing such comprehensive knowledge.

And here I think we will leave them. We certainly cannot help them, and most of us cannot even follow them into the rarefied atmosphere of higher geochemistry.

Investigators for some time have been dubious about the "mystery" of Life. They have studied and analysed it and have found that it consists of nothing but molecules, some simple, some complex, just as is ordinary inert matter. They have even been able to synthesise some of the animal products. They will probably never be able fully to understand the amazingly intricate constituents and reactions in, say, a human body, but they will get to know more and more, and actually there is no secret about it; it is merely complicated. Single-celled organisms present less difficulty but all that is found in a single-celled organism or a man is merely a certain arrangement of molecules. The whole set-up, in fact, is matter. So, say many, Life is Matter.

Of course it is. Life has got to use matter or, to us, it would not be life, it would be nothingness. Our senses can detect only material things. Life unhampered by matter may be teeming around us but we can have no contact with it. There may be ghosts, but if there are they can take no shapes and we cannot see or hear or feel them, though an overstimulated brain may imagine them.

Yet looked at in another way life is *not* matter; it merely uses matter. Only when the sceptics have *created* life out of matter, life that can reproduce itself and start on an evolutionary journey, can we believe that life is merely an arrangement of molecules. Until then we must regard it as a mystery.

CHAPTER 2

SEA, LAND AND ICE

WHEN life came to the sea, the occasion was greeted with no fanfare of trumpets. The sea remained as crystal clear and as apparently empty as it had ever been. And so it remained for five hundred million years.

What happened during this vast period of time? All that happened was that the sea began to contain bits of jelly and a number of forms similar in appearance to the organisms one sees in a drop of pond water under a microscope. Not much progress, one might think, for such a long time, especially when one considers that it took no less time for the first life to change into one of these bits of jelly than it took for the bit of jelly to develop into a man. Indeed, it probably took longer.

But we must not be deceived by smallness. The first life developed—in time—into a single cell, and a single cell is an incredibly complicated structure. It is really not very surprising that it took as long a time to make a small multicellular organism as it took to make a man from the small organism. Both are miracles.

It is fossils—those invaluable scribes—that tell us by their very absence that no life other than very small and soft could have been present until shortly before the Cambrian period (500 million B.C.). At the beginning of the Cambrian period, however, fossils of snails, sponges, trilobites, graptolites, sidneyias, etc., began to be laid down and the book of recorded life had started.

Without the evidence of fossils how do we know that life was present in the sea for 500 million years before the first fossils ap-

peared? In other words, how do we know when life first came?
Roughly speaking, in any examination, there are two forms of evidence, visual and circumstantial. Fossils represent the visual evidence, deduction supplies the circumstantial evidence, and with the
knowledge we possess today of embryology and comparative anatomy it has been deduced that life must have first appeared towards
the beginning of the Proterozoic era, about 1,000 million years ago.

No doubt, previous to fossil-recorded history, evolution and selection were just as busy as they are today amongst their small and
often filter-passing organisms, but those remote times and that
minute life are a special study for which there is no room in a
general survey. At the beginning of the Cambrian period larger and
harder forms of life were appearing and thereafter the fossils began
to work scribbling their story on the rocks.

This planet is mostly sea, and there is nothing really to prevent
it being *all* sea. A little natural levelling of land would do it. Even
as it is, only 29 per cent of our globe is land.

The sea swells around us and there are no restrictions or real
barriers to it. It knows nothing of the isolation of islands and continents, and its inhabitants have a freedom of movement equalled
only by the birds in the world of air above. Indeed, both birds and
fishes often make journeys of incredible distances. The sea, in spite
of its frequent fury on the surface, is a region of complete calm.
Under the first few feet perpetual stillness and quietness reign. We
on land, in a way, live also in a sea comparable with the ocean. It
is a sea of air deeper than any waters and we, too, like the bottom
fishes, are subjected to great pressure. There is this difference, however, between the sea of air and that of water. We, though we live
at the very bottom of our sea, are frequently subjected to the discomfort and danger of moving currents of the medium in which
we live. At any moment a hurricane may burst upon us. This does
not happen in the sea of water—a bottom current there, moving,
say, at ninety miles an hour would indeed cause chaos. The dis-
comforts of rain, snow, sleet, hail, are also unknown to the sea
dwellers; they have their vital element always around them and do
not need to be uncomfortably squirted and sprayed with it.

For this and other reasons which will appear shortly it is a won-

der that any sea dweller ever had the urge to transfer itself to land.
Actually none of them did have the urge, they were pushed there.

In saying that the ocean underneath the surface is always quiet
we do not mean that there is no movement of the water. Owing to
the fact that cold water is heavier than warm water and salt water
is heavier than less salty water (and to winds and the rotation of
the earth), there is nearly always movement in the sea. Cold water
from the poles creeps along the ocean bottom towards the equator,
moving perhaps at a speed of a mile a day; other currents flow here
and there, all caused by earth rotation, temperature or salt content,
apart perhaps from the slight influence of the tides. In fact, the sea
at any depth is restless, but this restlessness is not felt by the inhabitants any more than we on land feel the upward current of
warmed air on an utterly still summer day—we only realise it when
we see baby spiders shooting up under their parachutes as if they
were on elevators. The swiftest currents in the sea are the surface
currents, particularly the Gulf Stream as it hurtles on its amazingly
long journey attaining at times a speed of four miles per hour and
more.

Very important to life in the sea are the tides—the strange phe-
nomena that twice daily, or in some parts once daily, send the
ocean swirling against the base of rocks, only to retreat in a few
hours leaving a bare expanse of sand. This pleases holiday-makers
and adds to the infinite variety of the sea, yet until towards the
end of the seventeenth century no one had the faintest idea what
caused this so familiar and inevitable rhythm, though, as usual,
many explanations, including the breathing of sea monsters, were
given. It was Newton who attributed the phenomenon to gravity,
and now any seaside holiday-maker will tell us that the tides are
caused by the pull of the moon. Some will tell us that the neap
tides—the tides that only come up about half-way—occur when the
moon is in its first or third quarter, and that the spring tides—that
come the full distance and sometimes, when assisted by the wind or
surges, cause havoc on land—occur when the moon is new or full.
This is because both sun and moon pull at the earth. The pull of
the moon, the moon being much closer, is naturally greater than
that of the sun. When the moon is new or full it is more or less in
a straight line with the sun on one side of the earth and the two

pull together, and we get spring tides. When the moon is in its first or third quarter the sun exercises an opposing pull, and we get our neap tides.

This very simple explanation, like most simple explanations, is correct so far as it goes, but does not explain everything. Why do tides vary so much in time and speed and volume? Why in some parts of the world are there two tides a day, in others only one, and in others hardly any at all? In Tahiti the tide never rises more than a foot, in other places it rises forty feet. The reason is that the bed of the ocean is like the surface of dry land and possesses its own underwater mountains, valleys, and plains. The mountains and valleys set up their own swirls and circular currents which may oppose the advance of the tide or, if in conjunction, increase it. Only if the bottom of the sea were universally flat, would we get tides that regulated themselves on a fixed pattern.

Nevertheless, man has managed to tabulate the times of the tides for almost every part of the world far in advance, but this has involved much study and expert knowledge. How then can some fishes without study or training, but most decidedly with expert knowledge, forecast the various and ever-changing tides? Yet they do, and possibly with more accuracy than men. For some species of fishes have to lay their eggs in sand that *must* be dry for some weeks while the eggs develop and hatch. These fishes, with their eggs all ready within, wait until the day of the full spring tide and at the critical moment (with only a margin of time to spare) swim to the tide's limit, deposit their eggs in the sand, and then flee before disaster overtakes them. How they know not only the day of the month, but the hour of the day when the tide reaches its limit of expansion, is to me a mystery. So the eggs lie fixed in the sand with no water to wash them away until the climax of the next spring tide. For this tide the little fishes, hatched and buried in the sand, are waiting, and when it comes they go back with it to the sea.

"Very well," some questioner may say, "perhaps it *is* always more or less quiet under the surface of the sea and I dare say the temperature is more equable there than on land, and certainly there are no blizzards or rainstorms. But think of the compensating comforts of land. I for one, like my food, and the land supplies us with abundance and variety. Almost everywhere pastures, meadows, cornfields can be put into cultivation, and where this has not been done there exist vast areas of natural grazing country. This richness of the land gives us an infinite variety of meat, milk, poultry, game, grain, bread, fruit. What has the sea to offer? I know there are sea plants growing at the bottom, often in great profusion and often presenting more colourful gardens than any found on land—for those who are able to go down and see them. *But* I am told on good authority that these plants are inedible and cannot grow at all at any depth greater than six hundred feet—and only a very small proportion of the sea bed is as shallow as that. Therefore, the area under the sea that can grow any plants is insignificant. The rest of the sea floor is a barren waste, an almost limitless expanse of mud or rock, less capable of sustaining life than the most arid desert known on land. What a poverty-stricken region—the sea!"

Now this questioner was quite right in all the points he raised except in thinking (first) that the sea plants were "growing" at the bottom. They are only anchored there and are quite incapable of deriving any nourishment from the rich "soil" below. And (second) he was wrong in thinking that these "plants" are of use as food to fish. Indeed, the majority of them—the sea-lilies, etc.—though they look like plants, are animals needing the same food as the fishes and, although static, almost as expert as the fishes in catching and swallowing prey.

In fact, few sea fishes can live on vegetables; they have to live on animal life. The sea contains no creatures like our cattle and sheep and game and rabbits that convert our land herbage into sustenance for themselves and meat for the meat-eaters. There would appear to be a fallacy here. No community, whether it is composed of fishes or any other animals, can "live on one another." There has to be a basic supply of food. And this supply must be continuous.

The supplier of food to the earth, both sea and land, is the sun. It is the one and only source of nourishment and on land we can see (externally) the whole process, the greening of plants by chlorophyll and then the growth, which is the conversion by the chlorophyll of water and mineral salts and gases into such complicated molecular arrangements as proteins, sugars, fats, and other things.

These plants, as I say, are our only food, though we often have to eat them second-hand in the form of meat, eggs, and milk. Many townsfolk sneer at the country. Dr. Johnson was one. "He who has seen one green field," he said, "has seen all green fields. Give me a walk down Fleet Street." Without these green fields, however, he would have been unable to take a walk down Fleet or any other street.

In fact, our fields, etc., are merely factories turning out large quantities of food of every kind, but without the smoke and ugliness of human factories.

And vegetables are the only food the fishes have, though they generally have to eat them second-hand, and more often third, fourth or tenth-hand.

The vegetables of the sea are also manufactured entirely by the sun with the aid of chlorophyll but, unlike the vegetables of land, they are microscopic and rootless. A tumbler dipped into the richest garden in the sea and looked at through the light would show only transparent water, coloured red or green, but nothing more, unless there were minute animals present.

Why should the meadows of the sea be composed of plants so minute? It is because they need the sun. Water is really as opaque as a stone wall. The only difference is it has to be thicker than a stone wall before it can shut out light. This can be seen in any tiled swimming bath when one looks at the bottom of the deep and shallow ends. At a certain depth in the sea—not *very* far beneath the surface—there comes a point (even if a tropical sun is shining fiercely overhead) when all rays of light are cut off and the surroundings are black as the inside of a tomb. Therefore, in order to be able to imbibe the rays of the sun all the chlorophyll-using plants of the sea have to be near the surface, and the closer they are to the surface the better they function. That is why not a trace of them (except their dormant spores) is ever found below about fifty feet.

But why should they have to be so small that most adult fishes cannot feed on them direct? Their smallness is essential to enable them to float near the surface. Larger organisms would sink slowly to the bottom. True, certain flagellates by lashing their tiny forms assist themselves to keep near the surface, but in the main only microscopicness can enable a body to keep permanently near the surface.

The surface plants of course *could* manage to be larger by going in for intricate arrangements such as air bladders, but there is no necessity for them to do this, they function perfectly well as they are. Indeed, they function better, for their very smallness enables them to be permeated by the sun and the necessary minerals, without any need for the intricate tubes and other internal impediments necessary to land plants.

Chief amongst these plants of the ocean are the diatoms. They comprise much more than half of all the rest. There are others including the flagellates and certain blue-green algæ. Incidentally, not all flagellates are as benevolent as those that help to manufacture vegetable food in the sea. There are certain flagellates (trypanosomes) that give sleeping sickness. There are flagellates in the sea, too, that sometimes, by an excess of breeding, foul the waters to such an extent that they poison large quantities of fish. These dinoflagellates caused what was called the "Red Tide" that came and went along the shores of western Florida in 1947. So abundant were they that a pint of water contained as many as 60 million flagellates, and the sea was red and felt slimy when rubbed between the fingers. It was death to other life and was estimated to have killed over 50 million fishes, many of which were washed up on the beaches, which reeked horribly with their decay. And there are flagellates that are half-plant, half-animal, that manufacture chlorophyll and build tissues that way but can also eat food like an animal, and are, in fact, both vegetable and animal.

As I say, the diatoms are considered the most important food plants in the sea. They are mostly enclosed in silicon and under the microscope look like crystals of different shapes—rods, discs, cylinders, some with spines and other adornments, and nearly all of them beautiful. They reproduce themselves by division. There are plants, however, even smaller, called nannoplankton, or ultraplankton, which easily pass through the finest silk, and recent opinion is that these, in certain warmer parts of the sea, may well be the most important of the food providers.

Diatoms prefer cold water (from freezing point to 10° C.) and do not flourish in warm seas, which accounts for the rich harvests of fish (and whales, etc.) in the regions approaching the poles.

Like the plants on land, the diatoms commence their chief growth in the spring, and come to their full abundance in early summer. A decline then sets in. The decline is due to the fact that they have exhausted the mineral salts held in solution by the water in which they float. They must wait for autumn and winter; for in winter the surface waters of the sea cool and sink, and the bottom waters rise slowly to take their place. These bottom waters have had no demands made on them and are rich in nitrogen and phosphorous salts. They now form the new surface layer in which the plants will flourish abundantly, though invisibly, in the spring.

Who eats the sea plants? This brings us to that long chain that stretches from the microscopic to monsters like sharks and blue marlins. The gardens of the sea feed the whole of the mighty sea populations, and incidentally supply a large amount of the food of man himself.

The creatures that feed on the green (often brown, or reddish, though still containing chlorophyll) plant organisms are often themselves but little larger—minute crustaceans, the larvæ of barnacles, prawns, mussels, sea-urchins, starfish, sea-lilies, sponges, jellyfish, shrimps and very many others. These are then eaten by slightly larger creatures, who themselves are eaten by still larger ones, and so on until it becomes the turn of the big fishes.

The small grazers in the gardens, together with the plants they graze on, are known as plankton, "that which floats."

Of all the creatures that graze direct on the diatoms and other plants and thus start the ball of sea provisioning rolling, far and away the most important is a crustacean called the copepod. Copepods vary in size, though all are minute, but the chief diatom-eaters (Calanus) are mostly microscopic. They have oar-like legs by which (assisted by their long antennæ) they can paddle along in a series of jerks. Their numbers are prodigious: any harbour holds about a hundred times more copepods than there are human beings on earth. Let the traveller on any ship, says Shipley in his book, *Hunting under the Microscope,* tie a loose piece of porous cloth under the cold water tap of a bath and leave the tap running. In a short time the cloth will become stained, and a microscope will show that this deposit is composed chiefly of the bodies of copepods. Investigators have found that in early summer (when the diatom crop is thickest) a haul of about ten minutes will bring up half a million copepods of different sizes.

The copepod is a voracious feeder and eats half its own weight in diatoms every day. This sounds a lot, and a horse that ate that much fodder would be an expensive animal to keep, but the com-

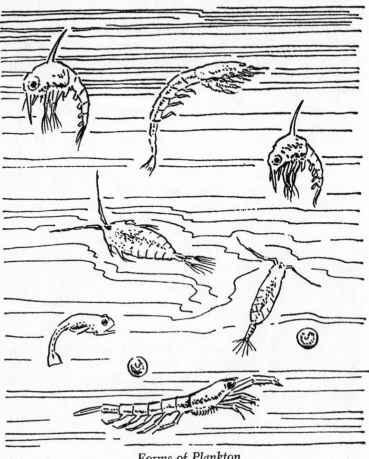

Forms of Plankton

mon shrew beats it, for that animal eats its own weight every few hours.

We have said that the plankton is distributed by progressive stages, the larger eating the smaller, so that the largest marine animals only get their plankton about tenth-hand. There are exceptions, however, and the largest creatures in the world are amongst them. The whalebone whales, for instance, and the basking sharks feed direct on plankton—which is tantamount to a full-grown man continuing to feed on the sieved preparations normally reserved for infants. A glance at a blue whale with its one hundred fifty tons of solid flesh and bone illustrates how nourishing and plentiful is the plankton in the sea. Amongst common fishes another exception (not quite the only one) is the herring, which, even when full grown, feeds on plankton only and on that diet produces shoals larger than those of any other fish, and flesh that is more nutritious and oily.

Special apparatuses are necessary for the larger creatures that feed on plankton—sieves that work automatically. The herring's food, for instance, might be compared with a saucepan of broth emptied into a bathful of water. It cannot pick out the particles, it must swim along with its mouth open and let the sieve (called "gill-rakers") do the sifting. The particles sifted out consist, in the herring's earlier stages, partly of diatoms themselves. Later, as the herring grows and its gill-rakers coarsen, the diatoms slip through, but small copepods are arrested. When adult, the larger copepods and other plankton-feeding organisms form the bulk of the food.

For those who like figures it has been calculated that a small copepod will have about one hundred twenty thousand diatoms in its stomach and a small herring will have about six thousand copepods. I do not altogether trust statistics, but going on, our information is that where they flourish there are about ten tons of diatoms per acre. Perhaps some farmer will tell us how that compares with his yield of hay. More reliable, perhaps, is the estimate of 400 million diatoms per cubic yard.

Man comes into the picture; he takes his share, and a big one, of the diatoms. He does this with every fish he eats and with a lot of his margarine. There are many chains that connect man with diatoms, but we will illustrate a very short and simple one. The herring makes migrations in search of diatom-feeding copepods (and diatoms grow best in cold water). The cod follows the herring, for the cod is the greatest eater of herring in the sea, or, possibly, out. The Labrador Stream, a current rather similar to the Gulf Stream, but cold instead of warm, flows (five hundred miles broad in places) from the Arctic down the New England coast to Cape Cod (where it meets and ducks under the Gulf Stream). This Labrador Stream is very rich in mineral salts, and by the time it passes Newfoundland and flows over the Grand Banks it has grown a large crop of diatoms which are, naturally, attended by a seething horde of copepods. And where the copepods are, there will the herrings be gathered together.

And where the herrings are, there will come the cod. Marked specimens show that cod travel from Iceland to Newfoundland.

Man, that great cod-eater, comes next. Off the Grand Banks the fishing fleets of many nations take a truly stupendous toll of cod.

So the direct chain here is: Diatom—Copepod—Herring—Cod—Man; and when man eats a herring (and he eats millions) the chain is one link shorter.

There is, of course, one further link in the chain—the very first—for the diatoms and other plants cannot exist without the necessary mineral salts. The salts of sodium, potassium, magnesium, calcium and some others are in unlimited supply everywhere in the sea, but, like plants on land, plants in the sea cannot grow or live without phosphates and nitrates. It is these two salts that regulate the whole food supply of the sea, and they are by no means present everywhere. They are distributed by currents and upwellings, and during temporary spells of shortage the diatoms disappear. Most of the migrations of fish are directly or indirectly due to the abundance or shortage of these salts in certain areas and at certain times.

As stated already, man's share in the provisions made by the sea plants is a big one. He eats more fish than meat; that is to say, the pastures of the sea sustain more stock for man than the pastures on land. Some races live entirely on fish. The bulk of the teeming peasant populations of China and Japan use only dried fish when they wish to supplement their rice. In America and Europe fish is a course in every restaurant, and fish suppers are the rule in many homes. In England "Fish and Chips" is almost a national institution, and a daily meal for millions. Consider, too, the amount of fish that is tinned. Glance at the shelves of any grocer; there will be tins of meat of course, but the bulk of the tinned food will be sardines, salmon, pilchard, herring, tunny, to say nothing of lobster, crab, prawns, shrimps, and their kind.

This is not all man owes to the sea. Fish-meal forms part of the food of poultry, pigs, and other stock. Crops are fertilised partly by fish manures, including guano, which is derived entirely from fish-eating birds.

Were the sea to present her account for the food supplied to man, it would indeed be a formidable bill.

And the sea gardens need no ploughing, harrowing, sowing, planting, manuring, weeding, harvesting, nor are they affected by gales, droughts, or floods. Why, indeed, did we ever leave the sea?

Land is only a part of the sea; it is a piece of sea floor that happens to be temporarily sticking out. Parts of certain continents, like South Africa, *may* never have been covered, but if so they are rare exceptions. It might be thought that land animals and plants have left the sea and have no longer any connection with her, but actually they are still almost as dependent on her as when they lived comfortably submerged beneath the surface. They rely on her for their food and water and also for a considerable amount of the oxygen they breathe.

Take man; he has colonised all dry land but he still lives on sea water. He is composed chiefly of water, and every drop he drinks is sea water carried to him from the sea by evaporation. If he ever does drink water that is not from the sea it is only a minute percentage that came from some erupting volcano.

The same applies to his food. No crops can be grown without rain (and irrigation from rivers or wells is the same thing). There can be no meat, bread, corn, or fruit without water sent by the sea. And a lot is needed. According to Furnas, a one-pound loaf of bread has used up two tons of water while the wheat it came from was growing, and an acre of corn uses three thousand tons.

It is often said of mariners that they have "salt in their blood," meaning that they love the sea. It is a foolish saying really, for we all have salt in our blood. Furthermore—and very significantly—we all have that proportion of salt in the fluids of our bodies that was present in the water of the sea at the supposed time our progenitors left it. And if this salt should be taken away we should die—quickly, but painfully. The sometimes severe pains of cramp after exercise are often due to too much salt leaving the body in sweat. Miners, stokers, etc., sweating much, and therefore drinking much water, used to be paralysed by agonising attacks of cramp. Nowadays salt is added to their drinking water, and these attacks are a thing of the past.

And the minerals in our blood are those that are also found dissolved in sea water. In short, we are all of the sea—salty.

So until we can grow grass and crops without water, and dehydrate our insides, and breathe less oxygen, we cannot really say that we have left the sea. We may toddle about here and there on our own, but the sea still bottle-feeds us.

Land and sea cannot be regarded as separate and unrelated. Most of the land we live on has been formed under the sea and is given to us only on temporary loan. And if the sea has a great influence on the land so does the land have an influence on the sea. It influences its volume, temperature, boundaries, nutrient content, aeration, and life. It creates and directs the course of ocean currents.

We think of the land as something fixed, as *terra firma*, but the crust that forms the earth's surface is not so firm as it appears. Were it possible to take an accelerated motion picture of any land over a long period with an exposure, say, every five hundred years we should view something that was as inconstant as the sea; quiet and flat at times, heaving gently at others, and at other times lashed into fury and throwing up waves higher than those ever attained by the sea.

There are two main movements of the land. The first is a rhythmic rise and fall, and the second is the eruption of mountain ranges.

The rising and falling of the land goes on continually and minor manifestations have occurred in recorded history and can even be noted in a man's lifetime. There are many once-pleasant villages now under the sea and many towns now miles inland that were once busy seaports. These changes are insignificant, but it is a different matter when Time (with a big T) is allowed full scope. For over geological periods the whole map of the world is changed.

The mountain-making upheavals are spasmodic, though it does so happen that they have occurred at more or less regular intervals of about 150 million years. There have been ten major revolutions (as they are called) in the earth's history.

When ranges such as the Alps were thrown up one imagines that it must have been a very impressive sight (and a very dangerous place for any animal life present). Actually, had a man been living there at the time he would have noticed nothing very much out of the ordinary. For the majority of the mountain ranges were formed slowly—slowly, that is, to our way of thinking but almost instantaneously from a geological point of view. Here again, an accelerated motion picture would show a spectacular scene—rocks as big as cathedrals rising from the bowels of the earth to thirty or forty thousand feet perhaps—we have no means of knowing the height of the mountains when they were new.

It is possible that we are today witnessing a mountain that is being "thrown up," for some geologists believe that Everest is still growing.

There are exceptions of course, and there have been great ranges formed amidst conditions that would have been spectacular even to a human onlooker. In the Proterozoic era, for instance, 1000 million years ago, a terrific upheaval of mountains occurred all over the world accompanied by earthquakes and volcanic eruptions on a scale never reached since (though there is no reason why such another display should not be in store). In places (in Canada and India particularly) millions of square miles were covered with molten lava to a depth of two miles and more. This lava is still very evident today.

The fate of great mountains is the fate of all mundane things— extinction. The grandeur of the Alps today is due to the fact that they are newcomers, parvenues amongst mountains, having been born a mere 40 million years ago in the Cenozoic era. For as soon as mountains are made the processes that are to level them start. Frost, rain, heat, ice, shave off their little every year. The "Everlasting Hills" are far from everlasting. Should you wish to see the grand old mountains of the far past you must seek out grassy, gently-

rounded hillocks, for unless it is on a plain, you will find them nowhere else.

So we have two processes, quite unconnected, the making of
mountain ranges which might be compared with the waves of the
sea in a hurricane, and the rise and fall of the land which might
be compared with the tides that occur regularly whether the sea is
calm or rough. There are times when the earth is low and flat, a
marshy affair of lagoons and low plains, and indeed these constitute
the normal aspect of the earth. Mountains and high plateaux are
the exception. We human beings live in one of those rare times
of earthen elevation. The usual condition of near flatness will return in due course, and by then, incidentally, the sea will have
taken possession of a quarter of what is now dry land. What the
human population is going to do when this happens I do not know;
they are crowded enough already.

When the earth is flat and sea covers most of the globe, the tem-
perature-regulating system works with the efficiency of a modern
central-heating plant. Among other currents, warm currents flow to
colder areas and currents going the other way complete the revolution. And the same with the air currents. There are no "blocks"
in the system in the shape of mountain chains or high plateaux.
Conditions are warm and moist all the year round. And these are
the normal conditions of earth but mankind has never experienced
them for he was born and still remains in one of the exceptional
hard periods.

When the land is elevated and mountains and high plateaux
cover a large portion, conditions are different: the water- and air-
heating system of the world goes wrong (just as domestic pipes
sometimes go wrong through blockages or in hard frost) and no
plumber can be called in to do anything about it. We get excessive
heat and bitter cold, gales, blizzards, aridity, and the rest.

These conditions affect life; species that have become too soft
die out, while others emerge that are better able to cope with cli-
matic severity. They also assist migration. The shallow seas are
drained off and poured into deeper basins. Isthmuses, long submerged, reappear and land bridges are made between countries and
continents.

Of great consequence to sea and land are the ice ages, those mysterious visitations of extreme cold that cause ice to creep like a rash and accumulate over a large portion of the globe.

Until recently the sun and earth combination was regarded as simple and obvious. There was the sun and there was the earth going round it and receiving its warmth. It had been going on a long time and doubtless (it was thought) in previous ages the sun had been hotter and the earth consequently warmer. However hot anything is, it will eventually cool, so it was taken for granted that the sun also was gradually cooling and would, in time, die out. After all, that was what the earth had done—red-hot and molten once, it was now cool and solid. Following this idea, you may remember, Wells wrote a book, *The Time Machine,* in which he pictured the earth of the future with the few remaining men shivering in glass houses while a dull-red sun, giving hardly any warmth, made its dreary circuit.

It is now known that things are not like that, and that the earth has an infinitely longer future than was once supposed. For the sun is not cooling; it is not just a white-hot thing, it is a living furnace with ample stocks of fuel in the shape of hydrogen. It is, in fact, a young sun with a long span of life before it, which means that the earth has a long span of life before it also. At the lowest estimate the earth will endure for twice the time of its existence up to the present.

Even so, the end of the world is a fixed and certain event. Cosmologists tell us that if our sun behaves like other suns (and there is no reason why it should not) it will in time come to the end of its supply of fuel. But it will *not* then begin to grow cold; on the contrary, for dynamic and other reasons, it will grow hotter. It will also grow larger. It will, in fact, swell like some monstrous balloon until it fills the space of the whole orbit of the earth, swallowing up and turning into incandescent gas Mercury, Venus, our own planet, and possibly Mars.

It *might* go even further. It might explode. If it did this it would not be following the example of most suns, and yet of a certain number. There is little doubt that the "Star of Bethlehem" was a finishing sun, a supernova, that exploded, and is even now suddenly

appearing as a brilliant new star to other planetary systems farther
away.

To go back: it was once thought that the sun and consequently
the earth were cooling. The fact that the fossils of tropical plants
had been found in both the polar regions merely endorsed this
idea. Knowledge of the various ice ages came, therefore, rather as
a shock. When, for instance, it was found that in the remote Paleozoic era, nearly 500 millions of years ago, ice, often several miles
thick had crept from the North and South Poles to the tropics, it
made people think. It did not force them to review their ideas
about a cooling sun, but they wondered how it could possibly have
come about. And we are still wondering today. For throughout
earth's history, ice ages have mysteriously come, refrigerating a
goodly portion of the globe and have, just as mysteriously, gone.

No adequate explanation has ever been given for the occurrence
of an ice age. The lifting of the land and the building of mountains makes, as we know, for cold conditions over large areas, but
this in itself is not sufficient. There are other explanations such as
the shutting out of the sun's rays by volcanic dust, sun-spots, and
many more, but none of these, even all together, suffice. It has even
been suggested that a period of extra *heat* from the sun could,
paradoxically, give rise to an ice age. It sounds absurd and we will
not go into the details, but there is a feasible explanation. In short,
with all our theories and guesses we are still in the dark. The only
certainty is that ice ages come and go, and that they do so at fairly
regular though widely separated periods.

The last ice age is still with us. It began about a million years
ago, when the glaciers spread slowly from the north. Quite four
times, these glaciers, like a tide, advanced and receded and the
animals and plants receded and advanced before them. The last
retreat—in America and Europe—began about ten thousand years
ago (which is getting very close to civilised human life) and is still
continuing. Whether it is merely a temporary retreat like the other
three, or whether it is the beginning of the final departure of the
present ice age, we do not know. At least the experts disagree about
it, which amounts to the same thing. In any case, it will not affect
us or our near progeny either way, so we can make our surmises
with that Olympic detachment that our short life gives us. There is

much to be said for a short life. Were we *really* long-lived we should have even more things to worry us about the future than we have now.

Man invariably takes everything for granted including the weather. We in temperate regions look upon our winters as the normal thing. We wrap sacking round the water pipes, lay in our stocks of coal, and get out our thick winter underclothes. Later we shovel away the snow from the paths and send Christmas cards depicting picturesque white landscapes and frozen village ponds. In the evening we sit closely round a roaring fire and pity those who are out in such weather and those who are unable to get coal. All this, as I say, we take for granted, a normal condition that occurs, with greater or lesser severity every year and always has done. Who of us pause to think that it is only because we are living in or just after one of those dreaded ice ages?

When an ice age *does* come about (which is extremely rarely) it always follows one of the periodic risings of the land whose effect is to send the waters of the sea back to their deeper basins. This, combined with the locking away of a huge amount of water in miles-deep ice-caps and glaciers, serves to uncover many tracts of sea bed that for ages had been deep in the ocean. Several times in the past, dripping from the sea, as it were, there have emerged broad bridges. America became connected with Africa, a broad band stretched from the Southern Continent to Australia, Gibraltar joined itself to North Africa, as did Italy farther east so that the Mediterranean Sea became two insignificant inland lakes, and in our own ice age Britain attached itself to the Continent by a land mass (the Thames and the Rhine became a single river whose bed can still be traced across the ocean), and a bridge formed between Ireland and Scotland.

This may appear to be merely of geological interest. It is more than that. The bridges made possible for animals (and plants) the colonisation of islands and continents that had been inaccessible before. Metaphorically speaking, they poured across the various bridges.

They did not realise it, but there was not much time. The bridges were merely drawbridges soon to be pulled up—or rather let down —so that although many got through, many arrived too late. Taking

Britain as a typical example, in addition to now extinct animals—
like the cave bear and the sabre-toothed tiger—hares, moles, deer,
foxes, lizards, snakes, and others got over the gangway, but others
did not.

Of these, all who went to Ireland had a difficult journey. They
had to go up to the north of Scotland, turn left, then south over
a narrow bridge. Again, many species made the journey and again
many did not, so that Ireland possesses fewer species than Britain.
The blue hare was a successful immigrant, but never really wanted
to go there. On arrival he found the place was not his line of country, but he delayed his departure too long. It was, however, ideal
country for the common hare, but the common hare did not find
the bridge in time. So when the sea again separated Ireland from
Scotland, the blue hare had not had time to get away and the
common hare had not had time to get into Ireland. Another animal
that missed, as it were, the bus, was the snake. Saint Patrick may
have done many things for Ireland, but he cannot be credited with
the absence of snakes in that country—if it *is* a credit, for snakes
are very useful creatures. Moles and roe deer also failed to cross
the bridge in time.

As the land sank and the ice partially melted, the jealous sea
gradually regained its lost territory and swept past the cliffs of Dover
onwards to the North Sea to make the map more or less what we
know today. Not quite; a large area held out against submersion
and became an island thickly crowded with men and other animals. These, unmindful of their fate, probably continued to lead
their normal lives until the ever-rising sea drowned the lot of them
and converted their island into one of the best of all fishing loca-
tions—the Dogger Bank.

As we have mentioned already, one of the effects of an ice age
is to deprive the sea of a not inconsiderable portion of its water
by locking it away in miles-deep ice-caps. We sigh often for
warmer weather and for no winters, conditions which obtained
when there were heated lagoons and lush vegetation near the South
Pole and when figs grew in Greenland. And it would no doubt
be very pleasant to have these conditions back again, but we do not

want them to come about too quickly. If all the present ice-caps were to melt in our generation so much water would be added to the ocean that London, New York, Tokyo and many of the largest cities in the world would be submerged fathoms deep.

CHAPTER 3

DEPARTURE FROM THE SEA

1

W HEN the Cambrian period arrived it was obvious (as fossils show us) that life had not spent its previous 500 million years in the sea marking time. It had split up into a bewildering multitude of species of all shapes, sizes and appearances, some protected by hard shells or plates, some mobile, some stationary, all of them small— though there did exist a gigantic monster called the giant trilobite that was eighteen inches long. With so many it is impossible to enumerate the species; there were primitive sponges, snails (the precursors of the molluscs including our present oysters and scallops), arrow-worms, jellyfish, shrimps, and hundreds of others with only Latin or Greek names. But above all, at this period flourished the trilobites, creatures that were a mixture in appearance of wood-lice, lobsters and crabs.

By the next period (the Ordovician—400 million B.C.) there had been changes and developments. Many forms had disappeared, and others taken their place. Amongst the new arrivals were clams, starfish, giant snails, corals, and sea-scorpions. The trilobites still flourished, but were on the point of handing over their numerical superiority to the nautiloids, newcomers with a straight cone-shaped shell, that were to develop later into octopuses and squids.

By the time of the Silurian period which followed (350 million B.C.) the trilobites were on the decline and the nautiloids in full sway. These had now crinkled their shell and coiled it spiral-fashion over their backs, while their heads with the long feelers were distinctly squid-like. Everywhere were to be found sea-scorpions (very lobsterish in looks and varying in size from an inch to six feet or more); sponges (after a serious decline in the Cambrian period) were back in full force, and earth worms had arrived. More interesting still: there were now fishes in the sea. Most of them were jawless, as the lamprey is today, and many of them were armoured outside, but as the period advanced came the acanthodians, fishes that had developed real jaws.

These changes, such as they were, had come about infinitely slowly, and no doubt there would be further slow changes, but, nevertheless, at that time, a permanent state of affairs seemed to have become established. Day followed day, and millions of years followed millions of years, and so apparently it would go on.

The *sea* was the home of life. There were barren rocks that rose from it into the nothingness called air that could be seen dimly from below the surface of the comfortable sea, but such rocks of course were quite incapable of supporting life.

That is how it must have appeared then. We know now that the seemingly impossible happened and that life *did* come out of the sea on to those rocks and actually live in a medium called air, which was entirely different from water. This is usually described as an invasion, but the process was more of an eviction than an invasion. The shallow waters near the coast and in the lagoons were becoming crowded with competing forms, and the weakest were already being pushed uncomfortably close to the margins, and even at times caught and stranded by retreating tides, and killed by the cruel air.

And so, as time went on, the seashore presented the curious spectacle of (amongst others) creatures with elongated and jointed fins that were able to flop about in the mud. But make no mistake, those elongated fins were not evolved so that the owner could venture farther from the sea, but so that he could get back to it again. In fact, sea-margin life at this period was not a brave array of indomitable adventurers mustering for invasion, but a bunch of misfits being turned out of paradise. And in due course, though they did not know it, they were going to be pushed farther—into the terrible conditions that obtained on the rocks and sand, right away from the mothering sea.

Consider the difficulties. An erudite scientist, had he been pres-

ent then (in the person, of course, of a fish, and living in the sea)
would have considered the idea as ridiculous as that of a child today asking its parents why they could not all go and live on the
moon. This, if he had the patience to do so, is how he might have
explained things to some finny inquirer of those times. "Firstly, my
friend," he would have said, "life could never breathe in air. It can
abstract only oxygen that is contained in water. Moreover, the body,
being 99 per cent water, could not exist except *in* water. In air it
would dry up in a matter of hours. Secondly, in the sea we have an
equitable temperature. The regions where we live can never freeze,
nor can they ever become unduly hot. Indeed, in the deeper parts
of our habitat the temperature rarely changes at all. On dry land
during the day, when the rocks were grilling in the sun, you would
get temperatures sufficient not only to kill you, but to roast you;
while at night you would experience (though not for long!) the
bitterest cold. And apart from day and night, there are the seasons.
Winter and summer mean little to us in our secure home; away
from the water both extremes would be fatal. Thirdly, as we talk,
you and I are stationary. You are a large creature, yet you can stop
still at any moment and go to sleep, suspended in comfort by the
water around you. In other words, in the water you weigh practically nothing.

"You ask what I mean by 'weigh'? You would soon find out, my
friend, what weight means if you were on land. There your body
would be cruelly pressed against the ground by a force called gravity.
All movement save for a few agonised twistings would be impossible.

"And that, I think, should be enough, for any one of these
factors would in itself make life impossible out of the water. But
let us go on. Let us for argument's sake suppose that the impossible
happens and that you leave the sea, breathe air, endure incredible
extremes of temperature, resist dessication, are not ruptured internally by the force of gravity. What are you going to do then?
At this moment, were you to see some food—a shrimp, another fish
perhaps—you would no doubt leave me abruptly, and by movements of your body shoot rapidly towards it, in the hope of securing
it. Were you on dry land, tell me, pray, firstly what you propose to
live on, and secondly how you are going to get to it? Your fins and

tail and body are admirably adapted to movement in the water, but not on land. So to the other drawbacks attendant on your hypothetical excursion to dry land, we must add the certainty of starvation. As I see it, to move at all on land would mean acquiring a completely different anatomical arrangement, pillars, let us say, capable not only of propping up the body against the force of gravity, but also able to progress by to and fro movements—one pillar being lifted to make a forward movement whilst the others remained stationary as they supported the swaying trunk. And so on. Assuming the impossible again, that you acquired such members, then your rate of progress would of necessity be so slow and clumsy as to be barely worth the effort.

"And lastly, having in this little intellectual game of ours got you on to land and enabled you to overcome a thousand impossibilities, and supposing you wish to make this new element your permanent home, how are you going to be able to breed?"

At this moment we will assume that the bored friend sees some prey and by means of superb anatomical co-ordination departs like an arrow to capture it. The scientist fish no doubt smiles and thinks of the lumbering progress it would have made on land by means of the imaginary supporting pillars. What would he have thought if he could have seen into the future and witnessed the movements of his friend's progeny on land—the racehorses, gazelles and cheetahs?

We know now that fishes *did* invade, or were pushed on to, dry land, and overcame all the objections cited by the finny scientist. And yet the scientist was right in a way, and in a few of his arguments, for the animals could never have *lived* on barren rocks and sand; all animal life, including ourselves, would to this day be flopping or crawling about on the sea and river margins, were it not for the sea flora, who were the real invaders of land. Animal life only followed after them like the hordes of Normans that entered Britain after the battle of Hastings was fought and won. Plants, not animals, conquered dry land.

Yet vegetable life had almost as hard a task as animal life. It, too, had to learn to live in air instead of water, to endure extremes of temperature, to get food and moisture from rocks and sand, and to support its own weight, for like the fishes it had no mechanism for doing this; hitherto it had been supported by the sea. And finally it had no means of breeding except in water.

Yet this ill-equipped army was the first of the invaders to march from the sea.

"March" it will be understood is a figure of speech. It was one of the slowest marches in history, and very unlike the almost lightning invasions that plants make today. Anyone who wishes to see a modern plant invasion need only make a clearing in a tropical jungle, or better still, just stay at home and cease to weed the vegetable plot, while the quick arrival of herbage on to new volcanic islands shows from what distances invading plants can operate.

But the invasion of the sea plants had to be accomplished by evolution—a slow process—and on to a barren surface. Indeed, a botanist today with all the wealth of modern land plants at his disposal might well find it difficult to find any able to cope with the harsh conditions that confronted the badly-equipped sea plants.

Nevertheless, once they had got a footing, the plants swept over the dry land with (from a geological point of view) amazing rapidity. Until the late Silurian period, about 330 million years ago, plant life had never, as it were, laid a foot on the land. Then the invasion began, and in only fifty million years plants covered most of the land, not with mere mosses and lichen, but with vegetation of every sort including massive trees—weird in appearance, but trees all the same.

Geological conditions probably gave impetus to the plant invasion. The long Ordovician period that followed the Cambrian was one of those times when the land was flat and low, the climate benign, and when water covered much of the earth. Indeed, at no time before or since has the earth been so flooded—though the sea contained less water than today; North America, for instance, was merely an island-studded sea. In the Silurian period that followed the Ordovician, the land rose again and mountains and valleys were formed, with the result that the sea retreated into deeper basins, leaving dry vast areas that had once been shallow seas. Plant life in such places had to adapt itself to new conditions or die. It not only adapted itself, but spread over the hills and valleys and over the face of the earth.

It was mainly the seaweeds that invaded the land, and its barrenness and aridity ought in itself to have been a sufficient deterrent. There are many harsh, stony and sandy deserts in the world today but none are so bare as the universal desert that dry land presented to the rashly advancing sea plants. Many gardeners today have a craze for rock-gardens wherein they grow plants specially adapted to a dry and stony habitat. But they do not just lay down a cartload of stones; they make pockets of soil. A certain amount of soil indeed is considered a pre-requisite for all green things. Yet no such thing as soil existed when the sea plants conquered land. It is plants themselves that make the "soil" by their dead roots and bodies. These, all matted together, provide anchorage, nutriment, a spongy base for holding water, and a cover for the rocks and sand which prevents their absorbing or giving off heat too quickly, and thus acts as a kind of thermostatic control against extremes of temperature.

Barren desert country defeats plants even today (or there would be no deserts), but there have been some remarkable reclamations. Many sand dunes, for instance, have been invaded and held fast by certain grasses. And well could the seaweeds have done with such grasses to help their advance. But grasses are a recent thing, an almost startling innovation in the vegetable world.

There seems to be a flaw here; if plant life cannot exist without a certain amount of dead plant life, how did the first plants ever advance? Probably the plants advanced slowly in relays of suicide squads, each squad providing the following squad with nutriment. And all these squads all the time were adapting themselves to life on land.

Needless to say, the harshest places remained untouched by plants for a long time, but they invaded in full force the moister and warmer places of the earth, including the plateau round the South Pole where no life can now live.

The deep-thinking scientific fish, had he turned his attention to plants, would probably have mentioned further difficulties in store for them. "You stand," he would have said, "erect in the sea. You also imbibe your food and minerals direct from the surrounding water. On land you could not stand upright but would lie hopelessly stranded, you could get no food, minerals or water, and finally you would dry up."

To stand upright may appear unnecessary; many plants, like

mosses and lichen and trailers, make no attempt to do so and get on fairly well, but in the bitter struggle that goes on at ground level the plant that can raise itself above the strangling undergrowth increases its chances of survival. In addition it gives its leaves more access to the sun and also weakens its rivals below by partially depriving them of that same vital commodity.

For a plant or a tree to grow upright is not so simple as it sounds. It is not just a question of a hard pillar fastened to the grounds by roots; the pillar has to be filled with complicated pipes working both ways. The leaves above take food from the air by the action of the sun and their chlorophyll, but they cannot do this on their own; they need also water and mineral salts in solution, including nitrogen, and these can only come from the ground, pumped by the roots through the trunk or stem up aloft—often at considerable pressure. How this pressure operates is not known even today. Capillary action in itself cannot account for the volume of liquid that gushes upwards, especially in spring. Although the leaves have stomata or openings, several thousand per leaf, that can be shut as required to suit conditions of temperatures and aridity, a normal-sized tree will evaporate into the air forty gallons of water a day during the summer period. (Trees, therefore, effect rainfall.)

The raw material, however, is only sent up to the leaves for processing. It has to come down again by other pipes to feed the trunk and the roots. So we get that gardeners' aphorism—the roots feed the leaves, and the leaves feed the roots.

So the seaweeds had not only to manufacture hard non-desiccating tissue before they could stand erect in air, they also had to manufacture supply pipes inside, and roots capable of selectively abstracting minerals and water from the soil. The roots in themselves were a big enough problem, for in the sea they had been merely anchors.

Incidentally, the bulk of the trees of this early period had no leaves as we know them. They had, however, green fleshy stems filled with chlorophyll and containing stomata which fulfilled the same purpose, if not so efficiently.

By the end of the Devonian period (275 million B.C.), large trees were flourishing in suitable places. It was incredibly quick work on the part of the seaweeds. True, the trees were, to our eyes, monstrosities, but they were trees, and we are using them as fuel today. It took a long time for them to develop into pines, oaks and beeches, but that mattered nothing, they had time to spare now, they had conquered the land.

Prehistoric Vegetation

So far I have said nothing at all about breeding, and yet that was the greatest of all the difficulties that confronted the vegetable emigrants from the sea.

Breeding, of course, is necessary for any invading force, vegetable or animal, that hopes to establish a foothold, but it is doubly necessary in the case of plants. Where a plant is born there it stays for the rest of its life. It can only advance in the persons of its progeny. In the sea, the plants, though static themselves, discharged their reproductive cells direct into the water in the shape of small but very lively swimmers. Such objects shed into the air would immediately dry and die. Here was another apparently insuperable problem, and the way the plants solved it is extremely interesting. Unfortunately, it is also extremely complicated, and a description would involve us in many pages of botany. It is sufficient here to say that plants gradually evolved spores that resisted desiccation, and were able to swim, when called upon to do so, in a microscopic film of water. That was enough to enable the plants to continue their advance, though the evolution of the modern seed, perfected in all respects for development on dry land, lay away in the future.

The plants had now laid down a carpet and a store of food ready for the animal tourists from the sea. And one would have thought that this was all that could have been expected of them. But they had another necessary part to play, without which no animals could have lived for long on land, and you and I would have been conspicuous today by our absence. The plants make oxygen, indeed they are practically the only source of that vital gas in the world. It is not present in any of the gases emanating from the earth or belched out by volcanoes. In the beginning what oxygen there was in the atmosphere must have been well-nigh used up by molten metals and permanently fixed in the form of oxides. Green plants had to renew it by breaking down carbon dioxide. This was slowly accomplished by the microscopic sea plants over millions of years and probably the atmosphere by the time of the Silurian period was much the same as it is today. But alone it would not have supported for long the myriads of animals that developed and spread over the land, had it not been kept oxygenated by the land plants as well as the sea plants. Plants, of course, are not mere distributors of oxygen. They need oxygen themselves and take it in at night, but they make more than they take, and all the oxygen we use and breathe (21 per cent of the atmosphere) has been made by sea and land plants. Without plants, even supposing we could get our food from some other source, we should be asphyxiated.

2

In the Devonian period (300 million B.C.) mild conditions returned, and the earth once more experienced the continuous summer that is its normal climate. This was accompanied (and of course caused by) a levelling of the land with a consequent flooding of the sea into lakes and lagoons and pools and broad rivers. These shallow waters had been for some time thickly populated by the ancestors of fishes. It was an ideal habitat in many ways, but for fishes there is always a risk in shallow waters. The oxygen content may get too low, or large areas may dry up, with, of course, a huge fish death roll. Slight movements of the land must have fairly frequently caused the drying-up of the lagoons and subsequent flushings. They must also frequently have sealed off the various lagoons from the sea, leaving them to be fed by streams and rivers, with the result that the inland waters lost much of their salt content, or became entirely fresh.

It was in these conditions that the fishes were evolved, i.e, in fresh or brackish waters, ór in the rivers themselves.

In unstable conditions no one main line of evolution can take place and the rapidly evolving fishes, flourishing at one moment and killed off the next, had to make some decision (the decision, of course, was made for them by selection). There was no future in staying, as they were, in those dangerous waters. So the bulk of the fishes returned to the sea. They were well fitted for this medium, for most of them had developed a self-regulating air bladder which enabled them to move about in comfort at any depth or pressure on one plane.

Those fishes that elected to stay in the danger zone were of two types, the lung fishes, and long, armoured types called the osteolepids.

Of these two, one group was to conquer land, and any betting would probably have been heavily on the lung fishes. This type had so evolved its air bladder that, to all intents and purposes, it could breathe. It was already independent of water. When the pools or lagoons dried up it burrowed in the sand or mud and could stay there almost indefinitely, even when the mud got as hard as a brick. But that is as far as it ever got. Time and time again in the history of evolution apparent perfection has proved to be a snare and a delusion, and has led species into blind alleys. In the case of the lung fishes it led them into the mud, and left them there, so that today what few species survive (there are six, one in South America, one in Australia, and four in Africa) are only found in muddy swamps.

The other group that stayed in the shallow waters gradually developed their gas bladders until they became practically independent of any de-oxygenation of the waters, and until they could exist out of the water for long periods. The process must have been a highly selective one entailing the extermination of species and individuals on a large scale, but there were always survivors to carry on.

We have seen plants occupy land and have now come to the second phase—the occupation by fishes. The fishes, it should be mentioned, were not the first animal emigrants from the sea; they had been forestalled by the insects, but they were the first vertebrate emigrants, and were the ancestors of all the reptiles, mammals and birds we see today.

For the rest of this chapter we will run over the main features of the fish invasion of land. It is a well-known story and perhaps seems out of place in a book on the sea, but I think this very spectacular development of what was once sea life demands some mention. Moreover, it serves as an introduction to the next chapter, for in some cases it was part of a circle and many species adapted themselves to land only to return to the sea and fill the waters with completely new types of marine animals.

You must picture now the Devonian period, 300 million years ago, a period of ever increasing warmth and humidity, knowing nothing of that season called winter. You must picture also land covered with pools, lagoons, and lush vegetation. And in these pools, shallow seas and lagoons you must picture the fishes we have last mentioned, condemned to death unless they could manage to surmount all the difficulties mentioned to his friend by the fish scientist.

These fish, as I say, gradually, and probably painfully, became able to live in air. But that availed them little if they still had to keep to their pools. What they needed now were legs, and to cut a long story short they *did* convert their fins into legs—of a sort. The gradual evolution of the fins into legs has not as yet been fully shown by fossil records. A bare idea may be got perhaps by studying the metamorphosis of a tadpole into a frog or examining those living fossils, the mud-hoppers, with their bent jointed fins that enable them to progress on land at a surprising speed. But make no mistake, the fishes' first legs did not enable them to hop, nor was there any hopping for a long time. All the fishes could do with their new legs was to push themselves laboriously along after the fashion of a man in a rowing boat trying to propel the boat by his oars over a sandbank. A better illustration of the first movements of fishes on land can be provided by yourself or if, like me, you are not as agile as you were, by some more active friend.

Lie face downwards on the floor, keep your legs together and quite straight and do not move them—for your legs now represent the fish's tail, and on land the tail was useless. Now propel yourself forward by pushing movements of your hands against the floor, *not bending the elbows at all*. If you do this you will indeed wonder why fishes left the sea—for that is how they left it.

Later the fishes developed a joint in the front fins. Continue your experiment, but you are now allowed to bend your arms. You will see how very much easier progress is this way—but it is not *that* much easier. In your experiment you changed from the unbent-arm method of propulsion to the bent-arm method in a matter of seconds. It took fishes millions of years to do that, and millions more to get any assistance from their hindquarters. It was a great "day" for them when they grew hind legs, a real advance—yet one of their progeny, the whale, who went through all this discomfort to get hind legs and to develop them afterwards, went to as much unconscious trouble to get rid of them when he returned to the sea.

Although it was so difficult, the fishes *were* able now to move on to land. Actually, the last thing they wished to do was to move on to land. They still regarded the land, clothed though it was in green, as hostile. And in a way they were right. The bare rocks and sand had been covered with vegetation, but provided no food, and fishes cannot eat green stuff direct. These leg-possessing fishes lived on animal life in the water and only used their legs to get from one pool to another when the animal life in their own pool had become exhausted: and to get to another pool when their own dried up. So it was to get to water, not to land, that fishes first grew legs.

Soon there appeared a strange creature moving about on the borders of the pools, a creature with a fish's head and a fish's tail, but walking on four short legs—Ichthyostega, the first amphibian.

The amphibians radiated out into a large number of forms, most of them bizarre, all of them clumsy with large bellies and short legs, many presenting forms somewhat similar to a crocodile—though the present-day crocodile has little connection with them.

The amphibians remained on the land for 50 million years. All this time they were tied to water. Doubtless they snatched up what insects they could on the land, but the bulk of their food supply still came from the water. They also needed water for breeding. The young of all amphibians, like the young of that present-day amphibian, the frog, must go through the tadpole stage.

The amphibians had a long inning. Nevertheless, they were lucky, for they lived through what must have seemed a period of never-ending warmth. But, as the Carboniferous period (250 million B.C.) with its moist greenhouse atmosphere came towards its close, another of those elevations of land took place. Mountains were formed, the world's central-heating system became disorganised, and an ice age (the worst—with the exception of the present one—the world has known) began to spread from the south almost up to the equator. The waters drained away from the lagoons, and arid conditions took the place of moist. Most of the vast assortment of amphibians were killed, though several lingered on until the Triassic period (175 million B.C.). Others returned to the sea.

Meanwhile, other forms had been evolving from the amphibians, small, mostly more agile forms, better fitted no doubt for existence in the changed conditions, but chiefly remarkable for one amazing characteristic—they laid their eggs on dry land.

These new forms were the reptiles, destined—if length of reign is any criterion—to be the most successful rulers the land has known. The reign of the mammals has been a long one, but there is a weary way to go yet before we can equal the 100 million years of occupation exercised by the reptiles.

The reptile egg was a most remarkable affair. The fact that we see something similar now in every grocer's shop does not alter the fact. The embryo still needed water but this was reduced to a minute drop enclosed in an inner amnion, and jealously guarded, as well as fed, by thick yolk, white and outer protective shell. Ponds were no longer necessary for the development of eggs. No human scientific discoveries have been as startling and as important as this. It marked the real conquest of dry land by animals.

There was, in fact, only one other step necessary before the erstwhile fishes could plant their flag, as it were, on the scene of battle, and take over the territory.

At a previous period some amphibian, or an early reptile, starving perhaps, must have eaten some of the green vegetation and managed to swallow it, and managed later even to get to like it. This first vegetarian was second in importance in the march from the sea only to the first layer of eggs on land. In fact, without either of them the invasion would have fizzled out and the onward march would have had to be discontinued. For *all* land dwellers, unless they are going to stick to the ponds and try to live on fishes and molluscs, must live on land vegetation. This applies equally, of course, to the carnivores, who are enabled to live only by the unwilling permission of the vegetarians. The vegetarians are the dictators; they dictate the size, anatomical make-up and habitat of the carnivores, who are merely in the position of camp followers.

The aridity of the Permian period (200 million B.C.) gave the reptiles the necessary push and more or less forced them to become accustomed to dry conditions at the start of their careers. They survived too, the new rigorous climate, cold-blooded though they were.

Naturally, a general anatomical improvement followed the new conditions. Legs became more efficient, straighter, longer and stronger, though bellies for the most part still tended to be unprepossessingly distended.

The Triassic and Jurassic followed the stern Permian age, and gradually the climate warmed and the seas flowed back, flooding the land including half of the present North America. The old type of reptiles began to die out and new and more active ones to take their place. These were the dinosaurs. They penetrated afield, even to the far north, to places that are frozen wastes or glacier-covered now, but then grew figs and semi-tropical fruits. They also divided out into creatures of every conceivable and inconceivable shape and size. Some walked on two legs, some on four, some wore huge bone shields and defensive armour studded with long spikes. Some grew

great frills along their backs that looked like the sails of eastern fishing vessels. Some grew heads like crocodiles, some like ducks, some
like ostriches, while others, though huge in body, had no heads at
all to speak of. They also developed their teeth. Some of these were
sheer murder—rows of six-inch-long daggers—others were suited for
making short work of any vegetation however tough and hard. (The
large, duck-faced vegetarian, Anatasaurus, probably set a record
amongst land animals by possessing no fewer than two thousand
teeth.)

Herbivorous and carnivorous dinosaurs were now sharply differ-
entiated.

The warm Jurassic period handed over to the still warmer Cretaceous period (100 million B.C.), and it was in this period that the
dinosaurs came to the full flower of their development. It seems
to be a natural law of evolution for species to increase in size and
there are many obvious advantages in size. Many of the dinosaurs
increased in the Cretaceous period to almost incredible proportions.
The climax came with the arrival of the Brontosaurus (though the
Gigantosaurus may possibly have been even bigger). This creature,
looking more like a mountain than an animal, weighed about forty
tons and measured seventy feet in length. Beside him an elephant
would look like a rabbit beside a cow. This is the largest creature
that has ever trodden the earth, and, unless the force of gravity
changes, it is probable that no larger creature *will* ever tread the
earth. Only in the sea is it possible for animals to grow larger.

This massive brute was entirely harmless and spent most of
its time browsing amongst the water vegetation, generally half-
submerged in order to ease the pull of gravity. Its very small pointed
teeth show that it could only have been able to deal with soft, lush
green stuff. It would have starved in a modern field. But there were
no grazers in those days, for there was no grass.

Far from harmless were the living engines of destruction that
preyed on Brontosaurus and its kind. As the vegetarian dinosaurs
increased in size so did the meat-eaters. They had to. To hold down
and kill and devour an animal of forty tons needs something exceptional in strength, size, and aggressive armourment. The unfortunate Brontosaurus and his colleagues met this exception in the
shape of Tyrannosaurus, an animal well on the way itself to being

Tyrannosaurus (foreground), *Anatosaurus* (background)

one of the largest of earth's inhabitants, but quite supreme in being the strongest and most destructive. This dreadful creature stood twenty feet high and measured fifty feet in length. No limbs before or since can have had the muscular strength of its enormous hind legs; a tall man would have come up to its knee. These legs ended not in ordinary feet, but talons capable of ripping open a rhinoceros or an elephant with one stroke. It stood on its two hind legs, dragging its massive tail behind. The forelegs were insignificant, but the head was very large and a cavernous mouth, lined top and bottom with long rows of six-inch stabbing teeth, completed the make-up of this nightmare. No creature that has ever lived on earth could have had any chance against it. Its favourite prey would seem to have been Triceratops, a monster itself, twenty feet in length and standing eight feet high at the stern, armed in front with three rhinoceros-like horns used for attack and defence, massive in build and possessing an armoured frill round its neck which served as a kind of shield for the body behind. A charge from this animal would undoubtedly have been able to disintegrate a modern house, yet against Tyrannosaurus it had no hope. Even a modern rifle would have made little impression. The bones of Tyrannosaurus were heavy, its brain minute, and any vital spot would have been hard to find in time, and even if found would probably not have had an immediate effect on its reptile constitution. And it must have been able to travel fast: even if its gait was lumbering its legs would have been able to cover a hundred yards in a few strides. So man may consider himself fortunate in that he did not live in the days when the dinosaurs held sway. Actually, this could never have come about, the dinosaurs had to go before man could be evolved.

The Cretaceous period wore on, and the dinosaurs came to the height of their power and peopled the world with creatures; small things, lumbering monsters, and active predatory giants. They also conquered the air. Some of their leathery-winged pterodactyls had the power of flight of a modern albatross, and one of them, at least, had a wing span of twenty-five feet which is fully twice that of the largest modern bird. They also produced a freak, a creature we call Archæopteryx that looked like a two-legged lizard bedecked with a few feathers along the arms and at the tail. It could not really fly; it could sail down from a tree, run along the ground, and from time to time make gliding leaps as it ran. And this is all it could do, though by now the pterodactyls were sailing like buzzards high in the heavens. Evolution is a strange thing, for the pterodactyls perished and the comical Archæopteryx begat every bird there is today. Incidentally, it was the feathers that did it, not so much in the first place as biters of air (the membranous wings of the pterodactyls

Pteranodon compared with an Albatross

were much better for that purpose), but as a quilt against the cold that came later, a cold the naked pterodactyls found themselves unable to withstand.

And now we find those original fishes that crawled so laboriously on to land complete masters of land and air. And of the sea also. An unconscious urge to return to the sea is probably latent in most land animals. The reptiles were no exception. Many went back. They not only went back to their old home, but they did very well there and became highly successful. They multiplied and increased in size until some of them threatened to become as great a scourge of the sea as Tyrannosaurus was of the land. Even the great Pteranodon, the flying pterodactyl we have mentioned, returned to all intents and purposes to the sea, for like some modern sea birds it probably never came back to land except to breed.

So there, at the close of the Cretaceous period were the all-conquering reptiles, paramount on earth, in the sea and in the air. Then they vanished. Like figures in a dream they were gone—the whole lot of them, the product of 100 million years of improvement and specialisation.

The vanishing of the reptiles (with a few insignificant exceptions) is now regarded as one of the world's mysteries. Their disappearance coincided more or less with the elevation of the land that took place at the close of the Cretaceous period (which also ended the vast Mesozoic era—175 to 70 million B.C.). Winters and aridity, so long forgotten, returned, and the waters, of course, drained back. So, until recently, it was generally assumed that the reptiles were simply killed off by an ice age.

There are, however, objections to that theory, and many countertheories. We will look at a few.

Aridity itself, to say nothing of cold, might well have killed off the large dinosaurs. An elephant eats many hundredweights of fodder a day and in Africa has to travel almost ceaselessly day and night to get that amount. (That is why the keeping and training of elephants in Africa has been and always will be commercially impossible except in a few favoured places. India has smaller elephants and regions of more concentrated fodder but even there the feeding of these monstrous beasts is not an easy matter.) Consider then, what a Brontosaurus must have needed in the way of fodder *per diem*. And the Brontosaurus was no traveller. It lay and browsed in and around ponds and must have found difficulty in moving its vast bulk about at all. The wonder is it ever got enough to eat. Vegetation in its day must have been very lush. And the same applies to all the other ponderous vegetarians of that time. A drying-up of the land with the consequent diminution of the vegetation required by these creatures would easily account for their extinction as would a drying and hardening of that vegetation. And *their* extinction, of course, would mean the extinction of Tyrannosaurus and the rest who preyed on them. One would have expected, however, many of the smaller forms, with their lesser demands on food, to have escaped. But very few did.

Did the unaccustomed cold kill them? The argument is that the reptiles had endured a severer ice age and got away with it very well. So they had, but those earlier forms were very different from

Dinosaurs compared with a Clydesdale cart horse: left to right—
Stegosaurus, Brontosaurus, Iguanodon Bernissartensis

the later dinosaurs who had evolved during many hot and steamy
years.

Another theory blames the *change* of vegetation. Flowering,
seed-bearing plants were taking the place of the old spore bearers.
It is suggested that the conservative dinosaurs could not bring themselves to eat this new herbage. That is possible, of course, but the
change of vegetation must have been very gradual. Moreover, the
dinosaurs had branched out in all directions and evolved species of
every kind, large and small. Not all would have been killed off by a
gradual evolution of plant life.

There are others who attribute the reptiles' disappearance to the
mammals. By the end of the Cretaceous period many small ancestral mammals must have been hiding in undergrowth or in the tops
of trees, leading hazardous lives amongst their dangerous neigh-
bors. It is suggested that these ate the larger eggs of the reptiles
and so reduced their numbers to vanishing point, much as egg-
eating rats may make difficult the preservation of pheasants. But the
smaller dinosaurs themselves must have been far more rapacious
than the sprinkling of mammals and undoubtedly ate up any eggs
they could find, and must have done so throughout most of the
dinosaur history. Indeed, the boot is probably on the other foot,
for the mammals themselves were egg-laying then and must have
had difficulty in preserving *their* eggs from the small dinosaurs. The
mammals later came into power *because* the reptiles disappeared,
but it is very unlikely that they had any hand themselves in that
disappearance.

It has been suggested, though very diffidently, that some disease swept through the reptiles' ranks and killed them off. If so, it must have been a very virulent and exceptional disease to kill off so many completely different genera.

(The dinosaurs, like many long-lived animals, were slow breeders, and this was probably a contributory cause of their decline when circumstances became adverse.)

All these theories, however, except perhaps the ice-age theory, are invalidated by the sea itself. For practically all the reptiles that had returned to the sea (and were doing very well there) by a strange coincidence (if it was a coincidence) perished at the same time as their land colleagues. This new ice age, of course, would make many parts of the ocean colder—and water dwellers are extremely susceptible to even small changes of temperature. Yet why, it has been asked, did not these sea reptiles migrate to the warmer equatorial waters? Migration, however, is instinctively acquired over long periods and the reptiles, knowing only warm waters, had probably had no occasion to acquire it. But this does not explain why those already in the warmer waters perished also.

The last theory we need mention is racial decay. "Racial senescence," as it is called, which means briefly that a race, like an individual, comes in due course to the end of its alloted span. The race dies, it is thought, from over-specialisation. And certainly many of the dinosaurs *were* over-specialised. The massive, armoured, shielded, Triceratops is an example. But others were not; they had entered no cul-de-sacs, and were keeping actively abreast of conditions. Did they, too, die of racial senescence? A creature that upsets this theory (except for those who think that an exception proves a rule) is a primitive reptile that joined the others in their excursion back to the sea. The giant sea turtle was specialised and armoured and, compared with most other sea reptiles, slow-moving. Yet he, practically alone of the sea reptiles, survived, and is with us now almost unchanged, and certainly shows no signs of dying from racial senescence. Its problem is to survive Guildhall and other banquets and land robbers of its eggs. The sea snake, Platurus, is another of the back-to-the-sea brigade that escaped the holocaust.

The fact remains, however, that the dinosaurs did suddenly die

Triceratops prorsus

out, so there must have been some reason for it. "Suddenly," of course, is used only in the geological sense and must not be taken literally. All we know is that their fossils disappeared after certain strata. The animals themselves may have lingered on in favoured places for many thousands or even millions of years.

As we have already said, the vertebrates were not the first invaders of land. The arthropoda (a phylum which includes spiders and insects) had planted their standard on the mainland some time before the fish expedition arrived. These were favoured for such an adventure. They had external skeletons which helped them to resist desiccation. By the time of the Carboniferous period they had multiplied and branched out into the innumerable flying and creeping forms that mankind has learned to detest and fear and are the only forms of life at the present time over which he has little or no control. Incidentally the insects were the first creatures to fly.

Their early start should have given them predominance and would have done so had they not made one mistake: they evolved a perfect breathing apparatus that took in air through a multitude of tubes that permeated through their bodies. This gave them that tireless activity that we know so well when we try to deal with flies or mosquitoes. Were it possible for a man to possess a similar breathing apparatus he would be able to sprint many miles at the speed he reserves now for a hundred yards (and do a hard job of work afterwards). This breathing system, however, though perfect for its possessors at the time they evolved it, had one limitation: it could only operate with creatures of their size. It broke down or functioned inadequately with any insect larger than, say, a Brazil nut. So while the one-time fishes were gradually growing larger until they culminated in beasts like Brontosaurus and Tyrannosaurus, the insects perforce kept to the sizes we know today. Indeed, certain dragonflies that flew about amongst the earliest of the insects were larger than any known now.

And this was very fortunate for the vertebrates. Had the insects been capable of developing on similar lines they would inevitably have destroyed all other life on land. There would have been no reptiles, no mammals, no men. Even as it is, they are presenting an ever-increasing threat. Midgets though they are, they have not abandoned their dream of dominance.

Woolly Rhinoceros

The third stage in the metamorphosis of the land-invading fishes was the appearance of the mammals. Their chief difference from the reptiles was that they were warm-blooded and brought forth their young alive. True, at the beginning they laid eggs, but the placental mammals with their complicated but more intimate system soon ousted them—though one or two egg-laying mammals still survive in Australia, that land of animal anachronisms. Together with the mammals in this third and (up to the present) last stage must be classed the birds, who succeeded equally well—indeed, if numbers are anything to go by, they succeeded even better, in spite of the fact that none ever managed to bring forth young alive.

As already noted, mammals had been on the world stage for a long time, but only as rare "wee, cowering, timorous beasties," chivvied and chased and leading miserable nocturnal existences. Eventual extermination would have seemed their only possible fate, but with the going of the dinosaurs they came into their unexpected own and entered the promised land.

It was practically an empty land they entered, but a green and pleasant one. The cold spell that had (or had not) destroyed the dinosaurs, soon passed, and warmth returned.

The history of the mammals was not very different from that of the reptiles. They multiplied and branched out and many forms increased in size.

Sizes in general continued to increase until, by the Oligocene period (30 million B.C.) monsters that rivalled even the great reptiles had come into being. Greatest of these was Baluchitherium whose skeleton was discovered by the American Museum expedition in Mongolia. It was of ponderous build and probably weighed twenty tons. It had a skull five feet long, a neck twelve feet long and forelegs measuring fourteen feet. This was the largest beast the mammals ever produced on land, though at sea, in the whale, they exceeded even the biggest efforts of the reptiles.

The era of mammals was the Cenozoic era which commenced 75 million years ago and is the era in which we are living now. Its periods were the Paleocene (75–60 B.C.), the Eocene (60–40 B.C.), the Oligocene (40–30 B.C.), the Miocene (30–10 B.C.), the Pliocene (10–1 B.C.), the Pleistocene (1 B.C.), and our present period, the Holocene or Recent, which has now reached its 10/15,000-year mark.

The Oligocene period presented a strange and weird assemblage of animals, uncouth and bizarre, most of which were shortly to die out, but (in America) one familiar figure would have been spotted sitting up or running and dodging about in its usual way—the rabbit. As a matter of fact, he was an old inhabitant even then and had been common in the previous Eocene period. He had no grass to eat—grass emerged from the forests where it had lurked and spread over the plains only in the Miocene period—but he was evidently as hard to keep under then, as he is now. (Myxomatosis is no new scourge to the rabbit; it has probably survived hundreds of such infestations.)

The end of the Oligocene period saw the extinction of many of the old-type mammals and the commencement of forms that gave rise later to our modern species. The Miocene period rolled by and the Pliocene took its place. Practically the whole of the Cenozoic era had been warm. London, for instance, and most of Europe had had tropical climates. Greenland was covered luxuriantly with plants. But as the Pliocene period drew to its close, an ominous chill came into the air, and the once warm winds developed a biting edge.

Meanwhile, the mammals had been developing, splitting up, and as usual, increasing in size, so that the period in which we live dawned on a world peopled by giants. Amongst them were bearlike sloths larger than elephants and able to rear up to a height of twenty feet, wolves six feet long, sabre-toothed tigers capable of pulling down a rhino, ferocious dogs the size of ponies, bears as big as hippos, and largest of all, of course, the mountainous mastodons and mammoths—creatures that in spite of their uncouth bulk had been highly successful and spread themselves over most of the land.

All this time the climate had been growing colder. Dense ice sheets were piling up in the north and soon these came creeping

Prehistoric mammals compared with a Clydesdale cart horse: left to right—*Mammoth, Smilodon, Megatherion*

southwards. The worst of all ice ages had commenced. The ice enveloped nearly a third of the world's surface and then began to withdraw. Four times it came and went. Its last retreat began roughly ten thousand years ago and is still continuing. Whether this is this ice age's final exit or a mere temporary withdrawal remains to be seen—though not by us.

The mammals stood up well to these severe conditions. Indeed, they continued to increase and thrive, moving backwards and forwards as the ice came and went. And then, some time about the beginning of the last ice retreat, the bulk of them, and practically all the large ones, perished with the dramatic suddenness that had marked the end of the dinosaurs. Here again no satisfactory explanation has been given. Since an ice age had given them birth it would seem improbable that similar conditions killed them, for they had already lived through a succession of ice advances and retreats. Man undoubtedly contributed to their destruction, but with the primitive weapons he possessed at that time he could hardly have taken a big part in the extermination of flourishing and widely-distributed species. Probably these large mammals perished not from one but from a combination of unfavourable conditions.

Will our present forms gradually increase in size like their predecessors? Certainly not while man holds sway. He largely controls animal selection now and is thus a big factor in evolution. Large forms are easy to locate and kill and appeal to man's craving for meat and his innate urge to destroy. They also use herbage he would like to preserve for his stock or, if carnivorous, deplete the stock itself. There is really no room now on this earth for man and any large "wild" creature. Small forms, such as rats and mice, though they cause him great loss, are protected by their smallness. Selection, as long as he is on the Board of Selectors, will favour continual reduction in size, so we may look forward in the future to mammals (with the exception of those bred and kept by man) not much larger than rabbits or rats.

The Biblical precept, the first shall be last and the last first, has often come true in evolution. In the mammalian era man (so some authorities maintain) began his career as a lemur-like creature and one of the smallest of the assembly both in body and brain. He has recently appeared as their supreme achievement. By walking on two legs he has been able to convert his forelegs into arms and hands capable of manipulations of almost incredible diversity and complexity. By the development of his hands alone he would be worthy of premier notice in any list of mammalian feats. And it was chiefly due to the development of his arms and hands, calling for ever-increasing alertness and concentration from the brain, that he evolved an exceptionally large brain from an exceptionally small one.

He attained his present position solely by virtue of brain and hands, for in all other respects he is backward. Although by the present miserable standards of the mammals he is a large animal, his physical prowess is poor. At any "open-to-all" sports meeting he would be outclassed in practically every event by some other species —even if he came from the U.S.S.R.!

We are, of course, mentioning man here purely as one of the mammals. This is no place to deal with his mental and spiritual attributes. And zoologically his arrival has been a misfortune. It has interrupted the interesting tempo of the development of the later mammals. He has exterminated and reduced species and interfered with the nice adjustment of the balance of nature. With children, and sometimes with adults, nightmares often take the form of some animal. In my own case, I remember, it was always a large bear. But what a nightmare to the other animals *man* must be! This latest product, this armed killer, walking on two hind legs like many of the dinosaurs, must have aroused more terror in them than ever did Tyrannosaurus in the other dinosaurs.

With man we come to the climax, to date, of the fishes' land adventure. In the following chapters we will deal with a few of the better-known animals that inhabit the sea today, but we have not yet finished with the mammals; we shall meet some of them again there. Evolution performed the amazing feat of converting them into perfectly adapted land animals and then went on to the almost equally amazing feat of converting them into sea dwellers once more.

Let us go back to the sea.

CHAPTER 4

RETURN TO THE SEA

1

A DOG with an admixture of wolf blood in its veins, brought up from puppyhood with human beings, will become as affectionate and devoted as any other dog. It will also display high intelligence. Yet in wolf country the day will probably come when it disappears and never returns. It has felt what is often termed the Call of the Wild and has joined a wolf pack, exchanging luxury for hardship. And anyone in Africa, or any country with similar conditions, who brings up the young of wild animals to a state of apparently complete domestication knows how liable they are to feel this Call of the Wild that forces them to go back to conditions only their ancestors have known and to which they have become ill-adapted.

There is another, more distant, call that has affected animal land life ever since it flopped ashore, and that is the Call of the Sea.

Throughout the long eras of the amphibians and reptiles there was a continual "back-to-the-sea" movement. Sacrificing their hard-won victory over the land, members of both these groups were from time to time reported missing. They had returned to the sea.

And in the sea many of them succeeded remarkably well, particularly those dinosaurs, the Plesiosaurs, the Mosasaurs, and the Ichthyosaurs, some of which became almost as large and murderous as their fearsome colleague, Tyrannosaurus, had become on land. Most remained reptilian in appearance, but Ichthyosaur practically converted itself into a fish, grew a large forked tail, and became almost indistinguishable from a shark. He might therefore be classed as the most thoroughgoing and successful of them all, had he not died out even before the others did. (We have already mentioned that anachronism, the turtle, who never made any change and yet, inexplicably, survived the holocaust both in the sea and on land.)

With the mammals, however, so much greater a time having passed since their ancestors came on land, one would have expected the urge to return to the sea to have died a natural death. It did not; apparently it remained as strong as ever.

Amongst the early mammals that spread over the land was one of which we have no record, or, at any rate, no record that connects it with its subsequent career. It was almost certainly small. Some (from vague anatomical deduction) believe it to have been bear-like, others think it must have been akin to the small ancestor of the horse. All we really know is that it was a small mammal and that it went out to sea, and, furthermore, to mid-ocean.

This was the whale, destined to become the largest of all animals. How it got to sea we do not know: enemies, competition on land, a growing taste for fish, even a desire to escape the persecution of insects, or just the Call of the Sea—we do not know. There is reason to suppose, however, that, unlike others that have returned, the whale did not journey to the sea via rivers and estuaries, but struck out, as it were, boldly from the shore. Such details do not matter; the main thing is that this mammal went to sea and stayed there.

Now to do this it had to undo most of the work that had converted it from a fish into what it was. It had to learn to live in the sea without being drowned, to discard the legs it had been at such pains to acquire, to grow a fish's tail or some other means of propulsion through water, to streamline its body, and finally (and one would imagine most difficult of all) to bring forth its single young, suckle it over a long period, wean it and care for it, all in the unusual medium of mid-ocean. Add to these difficulties the fact that the whale, being warm-blooded, had to devise some means of protecting itself and its infant from the cold of the Arctic and Antarctic, had to swim vast distances to breed—often from polar seas to warm waters near the Equator—and return the same distance to feed, and had to be able to dive several hundred fathoms and back to the surface in one breath.

And somehow these things came about. For all its size (often attaining one hundred feet in length and one hundred fifty tons in weight—the weight, that is, of about fifty Indian elephants) the whale became as beautifully streamlined as any fish. Its forelegs were converted into two fins, the hind legs disappeared altogether, the neck was dispensed with and the huge head became joined to the body so that the two formed a smooth curve. The external ear vanished, and even the male penis was made retractile. Incidentally, though the hind legs disappeared externally, in some instances, very small vestiges of their bones are still to be found embedded in the flesh at the hinder end of the animal. Finally, to complete the transformation, the whale became hairless. This might not seem to have been advisable in view of the importance of protection from cold, but hair, however thick, is little protection against the cold of *water*. In water any overcoat must be inside the animal and must itself be protected by a waterproof outer covering. True, the seal tribes retained their hair, but then they had to spend long periods on land.

Like a fish, the whale had to have a tail, and grew one. There was, however, a difference between its tail and that of a shark, or other fish—the flukes were set on a horizontal plane, whereas those of fishes are on a vertical plane. The whale swims, therefore, by straight up-and-down movements of its tail (combined with separate movements of the flukes) as opposed to the side-to-side figure-of-eight movements of the fish's tail, or rather of the fish's whole body. That this does not handicap it will be realised by all those who have seen those smaller whales we call dolphins and porpoises playing round a swiftly-moving liner. The reason for the unusual set of the flukes is probably because of the whale's frequent need to surface rapidly for air from great depths—a tail with horizontal flukes being obviously better adapted for this.

We have mentioned before that long-extinct dinosaur, the Ichthyosaur, another land animal that returned to the sea and became as successful in that element as the whale. Now this animal also grew a tail, but the tail was like that of the fishes, set vertically. Presumably a whale-like tail would have suited it better, but the

Ichthyosaurus (left), *Plesiosaurus Macrocephalus* (right)

Ichthyosaur was nearer to the fishes by a considerable length of time and therefore more inclined to revert, as it were, to type.

Startling though the external anatomical changes of the whale were in its transition from the land mammal, the internal changes were less. Outside it seemed to have changed into a fish, inside it remained a mammal, with a mammal's blood, flesh, guts, skeleton and lungs.

The whale breathes air, as we do, and yet it can remain submerged for thirty minutes or more. In other words it can "hold its breath" for half an hour—a feat definitely beyond us. So one is inclined to picture in the whale a specialised and elaborate breathing apparatus and lungs of enormous size. Surprisingly, we do not find this; we find lungs like our own but size for size, a little smaller. This seems a puzzle, and the answer to the puzzle is psychological rather than physical. We and the other land mammals die of suffocation in little more than a minute when oxygen is cut off by a pillow or water or any other means, and most of us know the agony of having to hold our breath even for a few seconds when under water. And yet when our supply of air is cut off there is enough oxygen in our lungs, blood, and tissues to last us almost as long as the whale, and even longer than the whale's normal time of immersion, which is about ten minutes.

The reason we die so quickly is simply that we panic—or rather the brain panics, which is the same thing. The brain controls the rate of respiration. When we take violent exercise our muscles need more oxygen and the brain supplies this oxygen to the muscles via the blood by making us gulp air rapidly into the lungs. That sounds reasonable, but where the brain slips up is in tackling the whole matter of respiration from the wrong end. Those parts of the brain that control mechanical movement have in the case of emergencies to be put in action by some urge that rings as it were an alarm bell. The more oxygen we breathe in, the more carbon dioxide we manufacture as a waste product. Now, strangely enough, what rings the alarm bell in the brain when we need oxygen is not shortage of oxygen (that *never* worries the brain!) but the presence of the waste product carbon dioxide. We pant when we exercise because the presence of a more than normal quantity of carbon dioxide has been communicated to the brain. The brain's response of increasing the respiration is for the purpose of getting rid of the carbon dioxide. When the brain finds that it cannot get rid of the carbon dioxide it is thrown into a panic, ceases to function, and the subject dies.

With whales the brain, no doubt after a long course of training, has lost its unnecessary sensitivity to carbon dioxide, and when the whale submerges and breathing ceases, allows this gas to accumulate without undue anxiety.

The whale, however, takes one safeguard which is beyond us. Its vital organs must always have an adequate supply of oxygen. So when it dives deeply it is able to shut off the supply of blood to the muscles and keep it in reserve for the more vital need of the central nervous system supplying the spinal cord and brain.

When the whale renews its breath it fills its lungs almost to full capacity. When a man breathes he takes in only about a quarter of his lung's capacity.

Apart from the supply of oxygen, the diving of the whale was once thought to present another puzzle. When a human diver goes down into the sea to any depth he has to be careful about his return. If he comes up quickly he will suffer from "bends" (or caisson sickness) which will give him pain and may paralyse or even kill him. As everyone knows now, this is due to the nitrogen in the lungs being absorbed by the blood under pressure. This does no harm at the time but when the pressure is relieved by the diver coming to the surface, the dissolved nitrogen comes out of solution and forms bubbles in the blood after the manner of an uncorked bottle of soda water. The damage depends on where the blood-stream takes these bubbles—if to the heart, of course, the diver dies. By bringing him to the surface gradually, with long and tedious waits at frequent intervals the nitrogen bubbles escape by degrees and are eliminated.

But the whale often dives very deeply. Gray states that a depth of eight hundred fathoms is not beyond the Greenland whale. Be that as it may, whales certainly dive not infrequently to five hundred feet, a depth that would cause the severest "bends" to a human diver if he rose from it in a hurry. But the whale *has* to come up quickly. He cannot afford to take several hours over the process like the diver does. Why then does the whale not get the "bends"?

This question, as I say, was once classed as a problem but a little thought soon solved it. It was, in fact, no problem at all. A human diver has air pumped to him during the whole of his submerged time. He takes part of a lung-full every few seconds. He is there-fore receiving a continuous supply of nitrogen. The whale makes its whole dive on one lung-full of air, so from first to last has very little nitrogen to cope with. Even so, it is prepared for any even-tuality. Before reaching the main nervous system (where any ni-trogen bubbles would be dangerous) the blood has to pass through a meshwork of small blood-vessels which would certainly trap any bubbles. At any rate, these small blood-vessels are there, though whether their function is to trap nitrogen bubbles we do not really know.

The whale is the only mammal that breeds right out at sea. It gives birth to a single infant once every two years. This seems to leave little margin for increase, but whales for long thrived and multiplied. That rate of breeding, in fact, is quite adequate, but is susceptible to interference. The coming of man with his genius for inventing weapons of slaughter has now upset the balance.

Generally speaking, animals born in the open must be mobile from birth. Thus hares, partridges and antelopes can run about almost as soon as they emerge, whereas rabbits, tree- or cliff-nesting birds, and human beings whose birthplace is more or less secure from enemies, have a long period of helplessness. Few places are more open than the sea, and the infant whale is born able to swim. As might be expected it is the largest infant known; a new-born blue whale measures twenty feet. It is suckled at the breast and is not weaned until it is six months old. For a long time after that —quite how long we do not know—it accompanies its mother and receives her care and attention.

The female whale's milk is rich. It is mostly cream and contains eight times more protein than human milk. It must be akin to the "Food of the Gods" imagined by H. G. Wells. Perhaps startling results might be obtained by feeding it to human babies, always provided it did not kill them. At any rate, the infant whale puts on eight times its original weight in six months. (This is another testimonial to the virtue of the sea's universal food—plankton.) The young whale, however, needs all the rich cream it can get. Like all young it must have warmth at first and the mother has probably journeyed to warm waters to give it birth. Shortly it must return with her to polar seas where it will need an inner overcoat of blubber to protect it.

As I have said, the whale is for a long period a devoted mother. It is not easy to study a whale's habit in the open sea but there is ample evidence about this. I will give an instance.

Whales, I must first say, are timid and nervous. They panic and flee when any of their number is attacked. In his book, *Whales of South-West France,* Fischer describes the harpooning of a large calf. It received three harpoons and the crew prepared to pull it in. But the mother rushed to the rescue. She made "unheard-of" attempts to free it, often trying to press it to her under her "fin," diving down with it, and trying in every way to drag it off in spite of the musket shots being fired at her. Eventually she did what might seem almost impossible and broke (with her tail) the three hempen ropes attached to the harpoons and freed her calf. But the calf was mortally hurt and died. The mother remained by its body until the next day although fired at continuously from the boat.

We first meet the ancestors of whales in fossils of the Eocene period (about 40 million B.C.). They have been given the name of Zeuglodonts and had reached about the half-way mark between the land mammal and present-day whales. They were smaller, of course, and were not yet really at home in the sea. They were less fish-like outwardly, and streamlining was a thing of the future. All were toothed fish-eaters; the plankton-feeding whales did not appear until the Oligocene period (roughly 20 million B.C.). All, however, had abandoned the land and bred at sea.

The whales of today are divided into two types—whalebone or non-toothed whales, and toothed whales. The whalebone whales are so called because they feed on plankton by means of sieves in their mouths made of long hanging brushes of "whalebone." The toothed whales feed chiefly on squids and fishes, and possess no whalebone.

We will deal with the whalebone whales (Mystacoceti) first. Whalebone, of course, has nothing whatever to do with the bones of a whale, which, except that they are somewhat hollow, are similar to the bones of any other mammal. In fact, it is not bone at all, it is more akin to the nails on our fingers. It grows only off the roof of the mouth in long plates (sometimes over three hundred in number) fairly broad and solid at the base and tapering off to divided hair-like shreds underneath. The whale moves through the plankton, or shrimp-congested "krill," opens its huge mouth, takes in a mouthful of water, and then lets the water run out. The whalebone hairs, assisted by an upward movement of the tongue, trap the teeming crustacea, which are then licked up and swallowed. (The throat of the plankton-feeding whale is almost ridiculously small.)

This may seem a fiddling way of getting a meal, but it is a very effective one. The plankton soup feeders are all much larger than the majority of their carnivorous colleagues, and food that can keep fat and well-nourished leviathans the size of the blue whales is not to be despised. (Two tons of plankton have been taken from the belly of a blue whale.) Would that we could send out our domestic animals to graze on such richness!

But whilst evolving this complicated but efficient feeding machine the whalebone whale was unknowingly storing up trouble for himself. The time came when ladies on land decided that the female form was not so divine that it could not be improved, and by a stroke of really bad luck for the whale someone perceived that the plates on the whale's palate might well be used as a stiffener of corsets, buttressing and moulding the female figure without too much discomfort. At one time whalebone reached a price of £2,000 a ton (which represents a vastly greater price today) and the whalebone whales were hunted mercilessly.

The whale was also hunted for the oil from its blubber, but later on it almost seemed that the whale was to get a reprieve, or at least a badly-needed breathing space. Women found that metal strips or elastic belts were adequate and less expensive correctors of their contours than whalebone, while petroleum, gas and electricity superseded whale oil for lighting and heating. (The price of whale oil fell from $2.75 per gallon in 1850 to 45 cents per gallon in 1900.) The slump in whalebone continues and its uses are now unimportant (it is chiefly used for brushes—and excellent brushes too!), but the whale was to be given no break. Some clever scientist found a way of converting oil into fats for a hungry world, and the chase was on again with full vigour and more deadly weapons, including radar-finding apparatus which makes any attempt at evasion on the part of a whale almost childish.

The whalebone whales include the blue whale, the fin whale, the humpback whale, the Sei whale, the Californian grey whale, and the right (or Greenland) whale. All these are or have been hunted for their oil. The only toothed whale of real commercial value is the sperm whale.

Before giving a short account of these whales, mention must be made of hunting in general, though so many interesting and exciting books have been written about it from the adventure point of view that it is unnecessary to go into the matter in detail.

Mankind has nearly always been short of oils and fats. That perhaps is why the story of the widow's cruse of oil has had such popular appeal, and the word "fat" has long been synonymous for richness. Phrases like "Fat acres," "Living on the fat of the land," and many more jump to mind. The recent lamentable groundnut scheme in Africa is another example of man's desperate need of fat.

Man must have been first introduced to the whale in the person of a stranded specimen and must have marvelled, first at its size, and then, more practically, at the amount of fat it carried. But stranded whales are a rare occurrence and if the mountain will not come to Mohammed, Mohammed must go to the mountain, and some bold man decided to go out and tackle a whale in its native element. When the first whale was captured we do not, of course, know, but it must have been quite early in known history, for when William the Conqueror invaded England whaling was well in its stride. And very exciting it must have been, with those primitive boats and equipment. Let those who are thrilled by witnessing the execution of whales by specially-fitted large steamers equipped with every device including explosives, try to imagine the thrill they would have got when capturing one from the cockleshell ships of olden days.

By the thirteenth century whaling was flourishing in Europe and revenues in various countries were being augmented by taxes on each whale brought in. In 1600 it was realised that whales teemed in the plankton-rich waters of the Arctic, though at that time little was known about plankton.

Thus began the Arctic whale fishing (the whale was then thought to be a fish) and it proved so successful that whales in those regions, numerous though they were, were practically exterminated. Consequently the whaling vessels have now concentrated on the Antarctic where the same process is taking place. Already many of the islands in the South Georgia region are marked by derelict, fallen-down buildings and the skeletons of whales, the sites of numerous once prosperous whaling stations, abandoned from lack of whales. Even the fish, long denied entry to the bays because of the pollution and poisoning caused by whale blood and offal are beginning cautiously to move back. Dimly aware, in the face of obvious evidence, that a goose that lays golden eggs *can* be killed, the nations now are making international agreements about whaling. At first it was not easy to get the necessary co-operation from all parties (before the last war, for instance, Japan refused to join in, hoping to profit from the restraint of others), but now all the nations have accepted international law, though it remains to be seen whether the whale will co-operate by keeping up its numbers against the legalised though still heavy slaughter.

The prize of the whalers is enormous. It is Big Money. Even just before World War II when the population of whales was at its lowest ebb, thirty-five thousand whales would be killed in a season. The war gave whales a *slight* chance of recovery—though not much at the rate of one infant every two years, and that infant a slow grower. A single modern factory whaling ship can, in one season, get whale oil enough to give a million people ample supplies of margarine for a year. There are many of these ships, so they can supply many millions of people with margarine for several years. They can also cut off the supply permanently if their official allowance is on too generous a scale, or if they break the agreement.

Let me now give some notes on a few of the better-known species.

The right, or Greenland whale must be given first place although it is now to all intents and purposes extinct. It was called "right" simply because it was the right species of whale for the old-time hunters to go after. It was so right that in those days (not, alas, very long ago) of pick and choose, no other whale was thought worth the hunting. Its oil was of finer quality than that of any other whale (except perhaps the sperm whale, but that whale carried no whalebone) and yielded more (thirty tons of oil from seventy tons of whale). It also carried more whalebone, and that of the finest quality. Its head was one-third the length of its body so there was room for immense quantities of that valuable commodity. (The plates ranged along the palate were about twelve feet long, ten inches wide at the base, and over four hundred in number.)

It teemed in immense numbers in the Arctic seas near the ice, so that it was once called "The Common Whale," or just "The Whale." How things have changed! It is now the rarest whale in the world. Its extermination (for so it can be called) would seem to have occurred in three phases. It was hunted first from about 1610 to 1720 in the waters round Spitsbergen and East Greenland; then, from 1720 to 1840, in the Baffin Bay areas, the hunting being largely in the hands of the Dutch; after that the destruction was completed by the Americans in the Bering Strait and Okhotsk Sea. America, in fact, put millions of dollars into the business and rarely has a venture been financed with more certainty of staggering profits. But here, as in the other parts of the Arctic previously, the inevitable end came in due course. Incidentally, hunters always seem surprised and aggrieved when after a long period of wholesale slaughter their quarry eludes them by allowing itself to be killed off.

The Greenland whale was peculiar in that it never migrated to warmer waters to breed. An evolutionary advance, one would think, in that it was able to breed where the food was and not half-starve in the plankton-rare warmer seas while giving birth to young. This asset, however, was cancelled by the coming of man, for a creature that keeps to one area is easier to annihilate.

The right whale was the most timid of all the whales. It had perhaps every reason to be so. If a small bird alighted on its back it would be thrown into a panic and dive as if the devil were after it.

The above remarks apply to the Greenland right whale, or Bow-head (*Balaena mystiectus*) of the Arctic. There are three other species that scientifically come under the heading of "Right"—the Black Right Whale (*Eubalaena glacealis*) of the North Atlantic; *Eubalaena sieboldi* of the North Pacific; and *Balaena australis* of the Southern Seas.

The humpback whale has a slight cavity in its back—hence the name. It is rather ungainly in outline also. It is chiefly distinguished from the others, however, by the great length of its pectoral fins. It is a sportive creature and delights in jumping out of the sea to a considerable height. On its descent its hundred tons or so cause no small splash. Male and female often lie together indulging in amorous pats with their fins. These caresses (which would knock an elephant head over heels) can be heard at sea for miles.

Its oil is poor and it carries comparatively little whalebone, so in the old days of plenty the whalers took little notice of it and the humpback whale remained fearless of them. When the pick-and-choose days closed down this whale came into the hunting list, and showed its appreciation of the fact by soon forsaking its old haunts and taking to new waters. By the same move it also showed a considerable amount of intelligence. Intelligence, however, is of little use to a whale when playing against man. Man has the loaded dice, and the once numerous humpback is getting rare.

The Californian grey whale used to abound off the coast of California whence it migrated in summer towards the Arctic. It was essentially an American whale, and American whalers went after it in no small way with the usual result. It was, in fact, definitely classed by authorities as extinct but recently one or two specimens have reappeared. Unlike the timid right whale it was often termed by hunters as "vicious" or "malicious." This opprobrium might better have been applied, one would think, to the hunters, for the viciousness and malice were only shown by mothers when defending their young. It was indeed a common practice for the whalers deliberately to harpoon infant whales and tow them to the shore where the harrowed attendant mothers could be destroyed in shoal water, making the handling of them afterwards easier.

This whale frequented coastal seas and was frequently stranded. Normally its huge weight when out of water gradually crushes a whale to death but the C. grey whale (an average specimen is about forty feet long) apparently waited unconcerned until refloated by the tide.

The sperm whale is widely distributed and still fairly common off the coast of Natal. It is an odd-looking creature and quite unmistakable, for the end of its head is not tapered but blunt like the end of a box or barrel. The toothed lower jaw (the upper jaw is not toothed) is set well back, superficially rather after the manner of a shark. The sperm whale, of course, gets its living in a different manner from the whalebone whales. It lives chiefly on large squids, and dives to great depths to get them. Incidentally, if Jonah ever was swallowed by a whale this is the species that swallowed him, for none other could have got him down whole.

Like that of the right whale, its head is fully one-third as long as its body. The head contains no whalebone but something which is now of much more value; a very fine oil (about a ton of it) called spermaceti, clear as olive oil when "tapped," which solidifies into a pure white fat of value for many purposes—medical, candles and cosmetics. Its blubber, too, is plentiful and of the best quality. This whale avoids the extreme polar regions. With all the whalebone whales the female is larger than the male. In the sperm the female is roughly only half the size of the male.

With its advantages (from man's point of view) it is to be wondered that the sperm whale still exists. It probably will not do so for very long and undoubtedly owes its survival up to the present to the fact that it carries no whalebone and was therefore not considered worth the effort of catching by the earlier whalers. Since then, however, it has been extensively hunted. America soon tumbled to its value and had a fleet of three hundred fifty ships specially equipped to hunt this whale. The fact that this fleet has long been disbanded tells its own tale.

Those interested (and who is not?) in the possibility of acquiring sudden wealth should hope for the continuance of the sperm whale. In this whale (and, for some reason, in this whale only) the hard part of a squid, the beak, sometimes sets up an irritation in the stomach. Nature seeks to isolate the source of irritation and surround it with a jellified pus, much as it does with us in the case of an abscess. Such a block of hardened pus is sometimes washed up on the shore. It is called ambergris. A lucky stroller on the beach of St. Helena once found a piece weighing 400 pounds with a value of £20,000. Some Norwegians found one of similar size inside the belly of a sperm whale and netted £27,000 for it, and in 1947 a piece weighing 340 pounds was taken from the intestines of a male sperm whale in the Antarctic (a large number of horny squid beaks were embedded in it). Such large pieces of ambergris are unusual, but even a small piece is worth, I am told, perhaps £500. Though it has a foul smell when taken direct from a whale, it is practically odourless when washed up and looks like a piece of blackish-brown clay. Few strollers on a beach would give it a second look, and one wonders how many have passed by such a gift from the sea without knowing it.

Sperm Whale

The sperm whale often voids ambergris naturally. The ambergris then floats in the water and may be washed up anywhere. Usually, of course, it is washed up on the tropical beaches of seas sperm whales frequent, but currents may take it anywhere, and it has been found on the coasts of Ireland and the Orkneys.

Ambergris has been known and prized for a long time, but it is only fairly recently (since 1724 to be exact) that its origin has been known. In the eleventh century it was thought to be bitumen that had been erupted from the deep sea. Later it was said to be gum from the roots of trees (hence the name "grey amber"). Even when it was found inside the bodies of sperm whales it was merely thought that they had swallowed it.

As I say, it has always been valuable, but not at first for scents. In the Middle Ages it was used to spice wine and food, and to cure various diseases including epilepsy. Wine spiced with ambergris must have been both expensive and (I should think) nasty. It is now used as a "fixer" for the more exotic perfumes. It has the property of chaining down the volatile scents. The bridge between an offensive smell and a sweet one is often less than we imagine.

The Sei whale, like the fin and blue whale, belongs to that group of whales called Rorquals. This whale is smaller than the other two both in size and numbers.

These Rorquals for centuries escaped the slaughter that was exterminating the others. This was not because man respected their attainments in superb streamlining and large size, nor even because they carried less whalebone; it was simply because he could not catch them. Until fairly recently whaling was done from small boats and the harpoon and line operated by hand. This served well enough with the other whales, but the fin and blue whales were far too fast for such treatment. The right whale was somewhat lighter than water, as were most of the others, but the fin and blue whales were considerably heavier and to draw up their monstrous bodies after death from deep water was impossible in most cases. There was also danger to the boat from their sinking, or at the least a probable loss of many fathoms of valuable rope. This situation changed with the invention of the harpoon-gun fired from large steamers equipped with steam winches for controlling the line.

Catching a blue whale then was simply a matter of pulling triggers and working levers.

The fin whale used to be known as the greyhound of the ocean. It is the second largest whale, frequently measuring eighty feet. Whaling now, with all its expensive equipment and vast amounts of money involved, depends to all intents and purposes on these two whales—the fin and the blue whale. Commercial whaling as we know it could not exist without them. One views the future with misgiving. When they go (and they are declining in numbers) an increased world population will face (amongst other shortages) a serious diminution in its fat supplies.

Apart from this, has the present generation the right (and men talk a lot about "rights" now) to annihilate the largest, and perhaps the most interesting animal evolution has ever produced? In destroying, for purposes of gain, the many species he has destroyed, man has nearly always been able to think up some excuse; the antelopes, etc., of Africa ate grass which ought to have been supporting herds of settlers' cattle, they damaged (unfenced) native gardens, they were aggressive and dangerous, they spread disease, they carried ticks, etc., etc. By no effort of his very powerful imagination can man think up anything against the commercial whales he is now killing off. They are gentle and timorous and they eat nothing that he can possibly pretend is of value to him. In short, they do him no harm and take nothing from him. The trouble is they *give* him too much.

Whales, however, are no longer largely wasted when killed. In the old days only the blubber (and whalebone) was used and the rest of the carcass left to rot. Now the rest of the carcass is ground up and used as fertiliser for the impoverished gardens of dry land. It is a pity the meat cannot be used for human food, for the world is almost as short of meat as it is of fat, but owing to difficulties of refrigeration and adequate transport only a negligible amount of whale meat gets to consumer countries. A fair amount got through to Europe in the time of World War II. It was not really liked, partly because it was strange. It had a certain oiliness (which could probably have been got rid of by suitable cooking). Several people in England liked it, but on the whole it did not "go down" except as a substitute. I ate it on occasion, but found it heavy and oily though scientists who have been on whaling expeditions tell me that whale meat from a *young* whale, well hung, is as good as any beef and has no perceptible oiliness. Anyway, the meat is no longer wasted as it was, but goes with the rest into the grinder to make fertilisers. The meat of a seal, though coming from a mammal more closely connected with the land than the whale, is a different matter. It is nutritious, I believe, but, according to many who have had to live on it for long periods, requires a young and cast-iron stomach to take it in, and after that it "returns" repeatedly.

With the toothed whales we are amongst the smaller fry. Except for the sperm whale there are no more leviathans. A length of thirty feet is about the limit for the rest—a size, incidentally, which would fill us with awe had we not met the blue whale and his shrimp-eating relatives. One or two of these others are still dignified in popular nomenclature by the name "whale" but the vast majority are simply classed as dolphins or porpoises. Amongst those few distinguished by the name "whale" are the bottle-nosed whale, the pilot whale, the white whale, and the killer whale.

The bottle-nosed whale has a bulging forehead but his head bears no resemblance to any bottle that I have seen. Whales of this species go about in small schools of between four and twelve and a spirit of real amity exists amongst them: they will never separate and, if one is wounded or harpooned, the others will only leave it when it is dead. Needless to say, whalers take the fullest advantage of this comradeship and having harpooned one of a school confidently look forward to killing all the others.

The pilot whale (twenty eight feet) is an all-black whale. (Most whales are white or whitish underneath) so that his other name is blackfish. This whale also has a bulging forehead which gives it an appearance of intelligence—though whether it is the deep thinker it looks is open to considerable doubt. Nearly all whale species have definite characteristics, characteristics that often lead to their undoing, and the pilot whale is no exception. His special weakness is a sheep-like, follow-my-leader gregariousness. It used to be common off the North American Atlantic coast and off Norway, the Faroes, Shetland and Orkney. This is the whale that used to be hunted by the inhabitants of these islands in special drives. When a school was sighted from the shore the islanders would jump into their boats and try to herd the school into one of the bays. Once they were in the bay one or two would be lanced and these, maddened with pain, would dash forward and become the leaders. The others would follow until most were more or less beached and could be dispatched with harpoons and lances.

The white whale. The word "white" as applied to an animal species is often completely unmeaning. The white rhinoceros of Africa, for instance, is just as black, if not blacker than the black rhinoceros. The white whale, however, when adult *is* white, and is the only whale to be so. It is about fifteen feet in length and lives solely in northerly seas. It is most abundant off America, both east and west coasts, and dives (so those who should know affirm) to very great depths to get its food, which consists mostly of crustacea and the inevitable squid.

Many whales make noises, often loud and weird noises. They are a vocal race. The white whale so excels that it has earned itself the name of the sea canary. Scammon tells us that when travelling on the surface it makes a noise like the lowing of an ox. There is a considerable difference between the sound produced by a canary and that produced by an ox, but it is, at any rate, well established that the white whale is vocal.

Since these whales are fairly reasonable in size, an attempt was once made to keep them in a tank. The first was brought from Labrador to London in 1877 and died almost as soon as it arrived. The second was caught the next year off Newfoundland and came to London without water in a packing-case full of seaweed. The journey took five weeks, and how the unfortunate creature managed to survive is a mystery. Yet it did, and swam about in its tank coming up at intervals to blow. It soon became very tame and liked the meals of fish that were given it. It died in a few weeks.

The killer whale is a large dolphin and I propose to talk about him later. (Just as a ship is referred to as "she," whalers always refer to a whale as "he.")

There are a large number of species of both porpoises and dolphins, but we need only consider one or two. First, however, will probably come the question, what is the difference between a porpoise and a dolphin? The answer is that there is none: all are toothed whales of the family Delphinidæ, but there is a kind of agreement that the larger members of this family (many of whom have beaks) shall be called dolphins, while the smaller beakless species shall be called porpoises.

The average voyager, however, need not worry about the different species. All he is likely to see is the so called Common Dolphin (*Delphinus delphis*) or the Common Porpoise (*Phocæna phocæna*). There is, of course, a certain *fish* that is also called a dolphin. It is brilliantly coloured and a great chaser of flying fish. It is remarkable not only for its colouring but also for its change of colouring after capture. When dying, these colours go all through the hues of the rainbow until death extinguishes them altogether. The scientific name of this fish dolphin is Coryphæna and, of course, it cannot be confused with the mammal dolphin which is black on top and white underneath and infinitely larger.

The frisky pack of creatures that passengers see from the deck, playing round a liner and cutting joyfully across its bows, cavorting, skylarking, jumping, splashing, racing and treating the ship with its many knots as if it were riding at anchor are, as I have said, common porpoises or common dolphins, and I have heard arguments amongst the onlookers about their identity. There can be no mistake. The dolphin has a beak like a bird's, the porpoise has none. The dolphin is about eight feet long, the porpoise is only half that length—it is, in fact, pretty well the sea's smallest whale.

Porpoise (above), *Dolphin* (below)

Both look very fish-like in the water. How un-fishlike they are inside was brought home to me once when I was travelling in a very small and very decrepit tramp steamer. Engine trouble halted us for several hours almost every day, and sometimes for days at a time. Fishing became the great pastime of certain members of the necessarily idle crew. Then they tried to get a porpoise. These, however, would never take a bait, and in the end one was harpooned and drawn on deck. Lying there, it looked like a fish, but when cut open its interior came almost as a shock. For inside were the bones, flesh, blood and organs of a newly slaughtered pig. When cooked the flesh was tough and tasted fishy. It had not, of course, been hung long enough, but even under the most correct treatment I do not think I could eat it. Yet in the days of Henry VIII it was a special and royal dish—which shows how tastes change, or how our stomachs have grown weaker.

All the dolphins and porpoises blow like the other whales, suckle their young on milk, and are just as devoted mothers.

Apart from the common dolphin two others are well known, the narwhal and the killer whale.

The narwhal frequents only the ice-bound waters of the Arctic seas. This peculiar-looking creature is known sometimes as the sea unicorn—for a single long-pointed spear juts out from its head, or rather its mouth. It is the male narwhal only that carries this spear and it comes about in this way: the various toothed whales have different assortments of teeth, some many, some few. The male narwhal, though it has vestigial teeth that never break the gum, has only two external teeth, one on each side. The right tooth, though starting in the normal way, ceases to develop and in many cases does not even break the surface of the boney ridge of the mouth, but the left keeps on growing as if it were unable to stop. It comes out of the mouth in the form of a spirally twisted ivory spear taking a direction on a plane with the narwhal's body, and continues developing until it has acquired a length of eight or nine feet. The female never grows any teeth at all; at least she does, but they serve no purpose for they remain embedded in the bone.

The function of this long spear (the narwhal itself is fifteen feet long, so its spear is more than half its body length) is not known, but many suggestions, of course, have been made. Some think it is used as an ice-breaker when the narwhal finds itself under frozen water (as must frequently happen). Some think it is used as a sort of skewer to impale and kill the prey (chiefly the squid family) on which the narwhal feeds—though the business of getting the prey off the long twisted skewer must present difficulties. Others say that it uses the tusk as a rake for stirring up the sea bed while searching for edible titbits.

Whatever its uses, if any, the spear cannot be indispensable, for the female gets on just as well without one. And she cannot rely on assistance from the male, because for long periods the sexes go about in separate schools.

Another suggestion is that the males, like knights of old charge their rivals with these lances for the possession of females. There are no records, however, of males having been found wounded in this manner.

But is it not idle to seek a use for *everything* in nature? Aberrations, in species, are always apt to occur and unless disqualified by natural selection will be allowed to remain. What use, for instance, are the whiskers on a man's face? Here also, as with the narwhal, the female gets on just as well without them. Nor can they seriously be thought to advance a man's chance of getting a mate.

Orca, the killer whale, the largest of the so-called dolphins, deserves more space than I can give him. This black-and-white combination of size, strength, speed, cunning, boldness, and ferocity, is the most dangerous proposition in the sea, not excepting the man-eating shark—though luckily its bulk prevents its frequenting shallow waters where, amongst bathers, it would be as a pike among minnows.

This whale, or dolphin, or large porpoise if you like, for it has no beak, is about thirty feet long and the infant at birth is about seven feet. It is only the male, however, that is thirty feet long, for the female killer (as with the sperm whale) is never more than half the size of her spouse, though equally ferocious. It follows then that the female produces an infant that at birth is nearly half her own length.

The jaws of the killer whale have great power and the many two-inch teeth interlock as they snap together. A bite from a Tyrannosaurus would hardly be more effective.

In shape the killer has attained perfection in streamlining; no tiny hump or hollow mars its perfect symmetry. Its speed must be outstanding, possibly surpassing that of the common dolphins and porpoises which some have put at forty knots.

The killer ranges all the seas of the world, but is mostly to be found in the Arctic and (especially) the Antarctic regions, where it finds its normal food in the shape of whales, seals, penguins, fish, etc. (In the Arctic, white whales and walruses figure frequently on its menu.) It hunts in packs of any number from two or three to about forty, and the members show aptitude for concerted tactics. These packs will attack the largest whale, some tearing out its tongue while others hold on to the tail in an attempt to render the huge creature powerless. It is the tongue they are after, for the thick blubber covering makes inaccessible the rest of the meat, and those who consider a mere tongue a rather small reward for such strenuous endeavours must remind themselves that the tongue of a large whale weighs from one and one half to two tons, which is the weight of a hippopotamus. Even so, this large quantity of succulent meat cannot go very far amongst them. Their appetite and capacity is enormous. It is not unusual to find the remains of a dozen seals in their stomachs and one specimen was found to contain the remains of fourteen seals and thirteen porpoises.

The records of Antarctic expeditions show the resource and cunning of these creatures. They will heave their heads out of the water against an ice floe and with beady eyes take stock of any life on its surface. On one occasion this calculating scrutiny showed them Ponting, of Scott's *Terra Nova* expedition, engaged in taking photographs. Shortly afterwards the floe began to break up as the killers (presumably with deliberate attempt) rose against it from underneath. By jumping quickly over the breaking portions and getting to land Ponting escaped, but—as they say in thrillers—only just. Nearly everyone, however, has read the accounts of Scott's last expedition, so there is no need to mention the other stories of killer whales there given.

Much as we may deplore the wickedness of this whale, there are zoologists and biologists who regret that its larger colleagues, so meek and inoffensive, have possessed none of its aggresive nature. Had they been given even a small portion, they would have been treated with more respect by man and thus been saved from early extinction.

What age do whales attain? No one knows. They apparently become sexually mature after three or four years, but that has no bearing on their length of life. The age of certain fresh-water fishes can be read from rings on the scales and it is known that carp and pike and others are still hale and hearty when over a hundred years. We have no data about sea animals, but an investigator claimed that annual marks were set up on the plates of whales that enabled its age to be determined. When it was found that, according to this, certain specimens were eight hundred years old the whole idea became a joke.

And yet . . . one wonders. A recent theory of biologists investigating growth and age is that deterioration (in other words, old age) only begins when growth stops completely; and the operative word here is "completely." So long as growth (cell division) takes place at all, old age is kept at bay.

We know that in the water there is no limit to the size of an animal. Freed from the fetters and strain of gravity it can attain almost any weight and dimensions. Stranded on land, a blue whale is crushed to death just as a man is crushed to death when buried under tons of masonry, but in the water it is as unhampered as a cloud in the sky. For this reason also (i.e. the neutralisation of the force of gravity by water) the various species of fishes, etc., never stop growing. After a certain period they only grow *very* slowly, but they *do* keep on growing and therefore do not really age.

That, at least, is the theory, and the theory would appear to be borne out by our present, very small, knowledge of fresh-water fishes. A fox, a man, a rabbit has not only a fixed span of life (if allowed to reach it), but also a fixed limit of size. That size is reached comparatively early and then remains put. An old man is no bigger than a young man. But take, for instance, the trout or the pike. I myself am a fisherman and have fished on and off for forty years. As a result of this lengthy period of collection by fly fishing, and other baser means, and judging entirely by my own experience and not what I had read, I would have put down the limit of size for a trout at about two and one half pounds. And so would thousands like me. We would, in fact, put down one pound as the normal size of a fully-grown trout. And were trout like rabbits and men this *would* be the normal size of a full-grown specimen. But the trout is not like rabbits and men; it is a fish, and lives in water. It knows little of the pull of gravity; never has to lift tired feet to trudge along and so never stops growing. If it escapes the hazards of all fish (which are great) it drifts back into some deep lake and in those calm waters grows *very* slowly every year. Thus we find trout weighing thirty pounds and more, and these old and heavy fish show no signs of senile decay when they fight for their liberty.

Rivers or even lakes, however, are a poor substitute for the sea. What a fresh-water fish gains from lack of gravitational pull it often loses in fighting perpetual strongly-flowing water. The sea gives more protection from gravity and has no strongly flowing rivers.

And so, if this theory of growth is correct, the creatures of the sea do not age. Most of them, of course, have little time to do so—their enemies see to that—but in the ocean depths there must be many monsters, squids, as it is rumoured, the size of tennis courts, creatures such as we catch ourselves from boats or rocks unrecognisable in the huge dimensions they have attained over the centuries—or over perhaps more than the centuries. Until the oceans are drained off we shall never know.

2

Long after the whale had left the land, another mammal felt the call of the sea and slid into the water. This mammal, the seal, made great progress and is now half-way towards becoming a whale itself. Its swimming powers are outstanding, though it is no match for those whales its own size, the dolphins and porpoises, and cannot cut playfully in front of a fast liner. It has already, when diving, schooled its brain to withhold oxygen from the muscles, keeping it for the vital central nervous system, and it can live permanently in mid-ocean, although it lives there now only for eight or nine months a year, and cannot breed there.

In the whale we see a finished product of evolution in the change-over of land to sea animal; in the seal we see an animal at the half-way stage. What the seal is going to turn into—whether it be another kind of whale or not—we do not know, but it is certainly not going to remain as it is. Many millions of years hence the fossil remains of seals will no doubt be of great importance, providing perhaps the scientists of those times with a sought-for "missing link" connecting some sea animal of their era with a land-living ancestor of the past.

Let us look at this half-way curiosity. It has a head and whiskers similar to those of a number of animals on land. (When swimming, its head just above the water looks remarkably like that of a spaniel.) It has a neck, though the neck is very thick and the seal is obviously in the process of streamlining its head to its trunk. It has converted its front legs into elbowed flippers that bear comparison with the elbowed fins of the fishes as they made their evolutionary journey the other way. For a swimming tail, the seal has joined its hind legs together so that they stick out behind and form the possible beginning of a tail such as a dolphin's. And ridiculous as these cemented legs look they are already extremely efficient in the water. It has retained its fur overcoat (for use when on land, no doubt), though it is now able, like the whale, to store an inner envelope of blubber.

In short, the seal, though only just out of the apprentice stage, is now fully qualified for a sea career. It has, in fact, so committed itself to the sea that it is barely able to move on land. It can flap and wobble along, and for a short distance move at quite a speed downhill, but the fact remains that on land it behaves like a wounded animal, paralysed from the hips downwards. Add to this that the seal does not feed on land and it is obvious that it would have forsaken the land long ago and made more marine improvements than it has, but for the fact that it has to come to land each year to breed. Until, like the whale, it can breed at sea it will always be a monstrosity, a half-and-half animal.

Will it ever be able to breed at sea? One sees no reason why it should not. Fossils have let us down rather badly over whales, but they started from the same starting point as the seals, and *they* breed at sea and obviously could not have done it straight away. Of course, breeding at sea will have to be *forced* on the seal. No animal changes its habits at all unless it is compelled to. In the seal this force may well come from its growing clumsiness on land. Most of its life is spent in the sea and evolution will therefore tend to make it more and more a creature of the sea, and the time may come when, like some palsied old man, it finds itself unable to climb on to the beach. It will then have to give birth to its children on the sea margin, and later in shallow water farther away, and later still, perhaps, at sea. The whole process might possibly take not much more than a few hundred thousand years.

There are many species of seals. The best known to us, in America and North Europe, are perhaps the grey seal and the common seal, the latter being much smaller than the former. Unlike some of their relatives, they do not congregate at breeding time into packed masses, nor do they make very long journeys. They haunt rocky, or sandy, coasts and bear their young there—or in caves. They are highly intelligent and inquisitive and, given any encouragement at all, over-trustful of human beings. They have no commercial value, no blubber to speak of, and no fur of the sort coveted by women. That is why they are common. They have, however, a bitter enemy in the fisherman, who claims that they seriously deplete the local fishing grounds and shoots them when they inquisitively follow boats. Refuting this, some experts have said that what they take is negligible, and so it is in comparison with what man takes, but seals certainly do eat a lot of fish. But why pick just on seals when there are dogfish and other sharks that eat infinitely more?

The fur seal (genus *Otaria*) once inhabited (during the breeding season) innumerable islands in the southern ocean in the North Pacific, and a large number of those in Arctic waters. These rookeries, as they are called, would be visited often by hundreds of thousands of seals. Many accounts have been written of the breeding habits of the fur seals in the past, and recently films have been taken of them. Since, however, their habits are rather similar to those of relatives such as sea elephants, which will be mentioned later, I will go over the ground again.

The date is any time before seal-skin coats became fashionable with ladies, and the place some deserted rocky island or beach, desolate and silent. In distant seas are millions of seals, scattered everywhere, eating their fill of the sea's rich stores, or sleeping, belly upwards, on the surface. All live individual, lonely lives, and enjoy it. Some are bulls of about seven hundred pounds, some are females half that size, and some are merely youngsters, who probably get together to a certain extent and have not travelled so far as their elders. Summer brings a feeling of unrest to the bulls, and from their various stations they begin to move purposefully through the water. They are bound for one pin-prick of an island on the map and may have to travel thousands of miles to get there. They may also have to round promontories or make other deviations. In fact, they have started on a journey that would demand considerable navigational knowledge even from a ship's master—with all his charts and instruments.

Later, back at the desolate island, an onlooker might see glistening, bobbing heads approaching through the surf. The first of the bulls are arriving.

Next day others come, then more and more.

All are sleek and fat in spite of their long journey. They heave themselves on to the beach and move ponderously up to the rocky mainland, where each flops down and noisily defies any of the others to come near him. They are selecting their breeding sites like so many prospectors in a gold rush staking their claims. There is much dispute over these small rocky patches of ground, and savage fights.

The island is silent no longer; the din is appalling; screams, roars, shouts, grunts, and hardly a claim-owner remains who is not ripped and gashed.

If it is like this when there are no females to stir up trouble, what is it going to be like when the unattached females arrive? And those females are now on their way.

A month passes during which the bulls mount guard over their small sites and never eat or drink or sleep. They do nothing but shout and fight.

In fact, they are looking considerably the worse for wear when another bunch of glistening heads is seen coming through the surf. The females have arrived and in one concerted roaring mob the bulls charge down to meet them like a brigade of ham-strung cavalry.

In spite of the fact that the cows are in an advanced stage of pregnancy, they get little consideration from the bulls who fight for them as a pack of hounds fight for pieces of meat.

Anon, out of the savage *mêlée* on the waterfront, emerge individual bulls dragging their prizes in the shape of buffeted and dazed females by their teeth. Each bull takes his bride to the site he has been at such pains for such a long time to reserve (and which may already have been taken by another) and there flings her down and resumes his guard duty—more necessary now than ever, for the screaming and fighting on the beach indicate that there are not enough females to go round and that soon frustrated males will come prowling about disobeying the tenth commandment. Even when all the females have arrived and been divided out there is little peace. Wifeless bulls are always causing trouble, while other bulls are not content with one wife if they think they can get more. Even the wives themselves are not above reproach and may desert their lord and master for another if they get the chance. So the bulls still have to be ever on guard and still have to go without food or sleep.

As a matter of fact, they go without food or sleep during the whole four months they are on the island. Many hibernating animals, of course, go without food for longer than that, but they are then in a comatose condition. To go without food during the period of an animal's greatest activity is a different matter and very remarkable, but not half so remarkable as going without sleep.

In about a week the females give birth to the young they conceived a year or so ago and things quiet down. Arguments may arise but on the whole there is no more brawling or fighting. This is not to say that there is silence again on the island; there is never any silence in a seal rookery.

The young thrive on their mother's rich milk and put on weight as rapidly as she loses it. Later, the mothers take the young for little trips to the sea, and teach them to swim, for although fashioned for swimming, seals have to learn the art.

Devoted at first, seal mothers have none of the whales' enduring love for their young, and soon they get away from them whenever they can. The fathers, or rather step-fathers, never take the slightest notice of them beyond a supercilious sniff.

So the pups, during the frequent absences of their mothers, get together in bands and romp and play together. After they are weaned (mothers and young have different ideas as to when this takes place) the mothers spend more time at sea, fishing perhaps, though by now most of the fish have taken the hint and gone to safer waters.

In addition to the bulls, the females, and the pups, there is a fourth class, the half-grown young that put out to sea and return each year to the island, like the others. These live in large bands farther inland, well away from the turbulent breeding bulls. They will not start breeding themselves until they are about fifteen years old.

The time of departure comes. Emaciated and starved, the bulls go wearily to the water and swim away. The cows and the young follow later, leaving the island, like a holiday camp at the end of summer, empty and quiet, and far from clean.

When the coarse outer hair of the fur seal is removed a sleek and beautiful inner fur is revealed. Women wanted these skins, and that was very nearly the end of the fur seal. America, however, saved the fur seal from complete extermination by severe restrictions on their slaughter, and there are now herds that breed on the Pribilof Islands (United States), the Lobos Islands (Uruguay), the Falkland Islands, and in the Australian region of the Antarctic. There is also a large market for fur-seal skins, but the killing of the seals is supervised by the various governments.

The sea lion (so well known to circus goers) is an ordinary seal and in spite of its name is as docile as all the others of its race. The Californian sea lion is the smallest of the genus. Highly intelligent, it is often kept in captivity and is a famous performer of "tricks." It is more nimble on land than the others and can use its tail to a certain extent to assist motion—a feat beyond the rest who find their tails mere encumbrances on shore.

The sea elephant is the giant of the family. Twenty feet long with the proportions of a hippopotamus, it is the most ponderous animal ever to flounder out of the sea. The cylinder-shaped proboscis over the fore end of its head may be responsible for the second name, but his bulk alone entitles him to be called an elephant. There are (or were) two species, a northern species once common as far south as the coast of California, now almost extinct, and a southern species that used to abound throughout the Antarctic up to the pack ice. It bred on the inhospitable islands of those regions but, unluckily for itself, was almost as great a commercial prize as a small whale. It could also be killed on land, doing away with the labour of towing and beaching the carcass. Add to this that, for all its size, it was as unaggressive as a dairy cow, and carried a vast quantity of excellent blubber, and it will be appreciated that it was a tempting prize.

The sea elephants come out of the sea and make their breeding sites some distance inland. The hunters prefer not to kill them there for that would entail having to drag the blubber to the beach, so they make them transport themselves to the slaughter grounds by raining blows (and worse) on their eyes and noses until, roaring with pain, the monsters leave their breeding patches and make back to the sea and are shot or otherwise killed just before they can enter that sanctuary.

Here, on the beach, the blubber is cut out and the rest of the carcass left to poison the atmosphere and water. Luckily the sea undertakes scavenging work even on land and her sea birds dispose of much offensive matter left on shore by man. It is a long, long time, however, before the air regains its untainted crispness, or the pollution disappears from the water.

Sea elephants would be extinct had not legislation been introduced. There are now at least two islands in the Antarctic where they breed, and slaughter is now conducted in an intelligent and more far-seeing way. If this restraint continues these unique and valuable animals may recover and increase.

The leopard seal or sea leopard (so called from its variegated coat) is an Antarctic seal living chiefly amongst the pack ice. Unlike its relatives it does not keep to a fish and squid diet and is a special foe of the penguins, who are often loathe to go into the water at all until they have pushed one of their members in to see if it is safe. The pushing is necessary because there is usually a shortage of volunteers.

The walrus is known to many chiefly on account of his famous walk and talk with the carpenter, and the illustrations to Lewis

Carroll's book usually depict him, with his bristly moustache and down-curved tusks, as a comically lugubrious individual. In the flesh, however, he is a formidable and ferocious-looking creature—but he *is* fond of oysters.

He lives only in the Arctic, generally in the vicinity of ice floes. He has developed no "tail" to speak of and is therefore a poor swimmer and quite unable to catch fish. He lives on shelled molluscs which he dredges from the sea bed with his tusks. Indeed, when opened up, the stomach of a walrus resembles the contents of an oyster-dredge. If a human being were to swallow only one of the thousands of sharp pieces of shell found in a walrus he would be whisked off to hospital. The down-curved tusks are about two feet long and are not used only for dredging: one of the most gentle and inoffensive of creatures, the walrus is, like many other gentle creatures, dangerous when roused and has the tusks and size and strength to inflict serious injuries. Its chief enemy (apart from man) is the polar bear, but even the polar bear would have little chance against a walrus were not the latter handicapped by its slow movements. Given the opportunity, walruses combine against an enemy, and when they do so polar bears remember other engagements elsewhere. The walrus also uses his tusks to hook on to the ice and draw himself up from the water. Both sexes have tusks.

There was a time when walruses were common both in the North Pacific and North Atlantic seas (the two species differed slightly), but that time has passed. For the sake of their oil, hide and ivory most of them have been killed off. Soon, like the other grotesque animals of Alice's dream, they will have disappeared.

Seals have been of use to mankind generally—even if only to provide women with expensive coats—but to certain races they have been as essential as the air they breathe. The Eskimos could never have lived in the Arctic without seals and walruses. For two thousand years those animals supplied them with practically all their necessities—food, clothes, blankets, oil for heating and lighting, needles from tusks, sewing thread from sinews, and boats. The number of seals and walruses killed by Eskimos has therefore been great. And yet it has really only amounted to a little thinning off. When man kills only for his own immediate wants he is part of Nature's

pattern and does no damage. It is when he kills for profit that he
becomes destructive.

Since the sea cow, owing to its name, is sometimes thought to
be some such animal as a sea lion or a sea elephant or a sea leopard
it might be as well to mention that it is nothing of the sort. It is
more whale- than seal-like and was once classed amongst the whales
(Cetacia), but it really has little resemblance either to a whale or a
seal.

The dugong, or sea cow, is interesting in many ways, not the
least of which is that it is probably the original mermaid. It has a
head and neck and shoulders similar proportionally to those of a
human being, and the flippers are, like human arms, situated at the
shoulders. So when it "stands" (as it often does) upright in the
water with half its body showing and its flippers folded across its
chest it may well—at a distance or in uncertain light—have a resemblance to a man or woman. Add to this that the udders are in almost
the same position as the human breasts and that the female clasps
her baby to her udder by one of her flippers, and one can well understand how the mermaid myth arose. In the early days, in the
Red Sea and elsewhere in that region, travellers to the East would
see dugongs from time to time and would spread the tale when they
got home. Those that go down to the sea in ships are expected to
see strange things, and what stranger than seeing a mermaid? The
old Spanish explorers called them "women-fish." I possess photographs of dugongs propped up in an erect position and they look
horribly human.

The dugong possesses a tail like a mermaid's too but lacks her
shapely waist.

It is not only greenhorns who are deceived. In the Bombay *Natural History Society Journal* it is related by S. H. Prater that in
1905 John Kahu, master of the cargo vessel S. S. *Samshon,* when
steaming in the Red Sea, saw what he thought were three human
beings standing waist-deep in the water. He took them to be survivors from a wreck, signalled to them, and made towards them.
But they did not answer his signals and when he got closer they
began to dive under the water, reappearing after some minutes.
Convinced by now that they were not human beings and did not

need rescuing, the captain went to the other extreme, got a rifle, and shot at the largest, hitting it, as it turned out later, in the neck.

I will finish off this tragic story in a moment. First we will have a look at sea cows in general.

Dugongs and manatees belong to the order Sirenia, so-called because of the fancied resemblance of these two and only members of the order to women. (Sirens, I need hardly explain, were those beautiful women of Greek mythology who lured mariners on to rocks and destruction by the sweetness of their song.)

The dugong lives solely by grazing on the seaweeds or grasses in the shallow reaches of the seas which it frequents. Hence perhaps the name "cow." It lacks the whale's or seal's ability to submerge itself for long periods and, so far as observers can judge, can only stay under water for about the same time as a human being. Some say for only a minute, others put it as long as five minutes. It keeps to warm waters and is found in the Red Sea, off the northwest coast of Australia, in some Indian waters, and off the east coast of Africa. I say "is found"; "was found," unfortunately, will be the correct tense very shortly.

Its flesh has always been reported to be outstanding. A delicious compromise between veal and pork, say some. Like the best beef steak, say others. Rather conflicting comparisons, but a taste is hard to describe. We can safely say, however, that it is good—much too good for the well-being of the dugong.

Sea cows are large creatures. The male is about nine feet long and weighs about three hundred pounds. Male and female mate, it is supposed, for life and are devoted to each other. The female produces a single young which is nursed at the breast and cared for over a long period. Both parents give their young an almost doting affection.

No more inoffensive a creature has ever been let loose in a savage world. It is easily taken, though the mother, if suckling her young, holds on to it with such tenacity that, according to native fishers, it takes several men to tear it from her.

It is disconcerting to learn that weeping is not the prerogative of women and children; the dugong also weeps when in distress. It possesses glands which secrete a watery fluid and which are connected to the eyes. Native fishers state that when a dugong is captured and bound tears literally pour from its eyes. This is confirmed by Dr. Deranigagala of the Colombo Museum who observed that a recently captured dugong wept frequently.

I will now finish off the story of Captain Kahu: the three dugongs (for such they were) made for a near-by island and managed to land on the rocky beach. The captain followed and sent men after them. They were found to be a male, a female and a calf. The wounded bull offered no resistance and was bound with ropes. The female also submitted to the same process but clung desperately to her calf. The three were put into a large tank on board where, in spite of everything, the mother continued to suckle her child. For food these rigid vegetarians were given fish—which is on a par with trying to feed cattle on rabbits. So in effect they got no food, yet they lived for two months, sustained no doubt by their blubber. Then the bull died, either from his wound or starvation or both. The female at once went frantic and searched for her mate in every corner of the narrow prison. Worse still, she abandoned her child so that it died in a few days, while she herself survived for only a few days after that.

Very closely related to the dugong and somewhat similar in appearance and habits is the manatee. It lives however on the opposite side of the world, being found in a few tropical South Atlantic regions off America and Africa (the two are different species, there is also a third species called the nailless manatee). Neither normally leaves the water, but while the dugong keeps to the sea the manatee often travels up rivers. In fact it has been found a thousand miles from sea water, and species have probably established themselves permanently in the upper reaches of such rivers as the Amazon and the Congo.

There was once another sea cow, infinitely larger than the dugong and inhabiting cold northern seas. This was Steller's Sea Cow (*Rhytina stelleri*). Steller, incidentally, was a German naturalist and he accompanied the famous Danish navigator, Vitus Behring, on his last voyage in 1741. Their boat came to grief and they were wrecked and stranded on an island, now known as Behring Island. Behring died a month later but Steller survived and existed on the harsh island for five months during which time he had every opportunity to study the hithertoo unknown sea cows which browsed and played in the Behring Sea and Straits in enormous numbers.

The dugong, you will remember, averages nine feet in length and weighs about three hundred pounds. Steller's Sea Cow was about thirty feet long and weighed about eight thousand pounds. Forty men, says Steller, were needed to drag the body of one of them on to the shore.

Numerous though they were, these sea cows of Steller had not long to live. Their days were numbered from the moment they were discovered, for they possessed three fatal qualities: they were trustful, inoffensive, and carried several tons of rich meat. They were completely exterminated by Russian sailors in a matter of twenty to thirty years.

The dugong and the manatee are also on the road to extinction, their sands are fast running out and for the same reasons that led to the obliteration of their larger relative. There are very few of them left now and it is doubtful if protection can be effective. It is a pity, for they are a unique type of mammal.

The origin of sea cows, like the origin of whales, is rather a mystery, but there are some indications. Like whales they came from some land animal but definitely not from the same animal that begot the whale. It was an animal that never went far from land as did the whale's bolder ancestor. In the embryonic stages the sea cow possesses a dense coat of hair which tends to show that it is descended from some furry animal, and fossil remains indicate that it was a four-legged animal. Resemblances in the brain, the

Manatee (left), *Dugong* (right)

structure of the heart, the position of the teats, and other anatomical similarities have led investigators to the conclusion that the
ancestor of the elephant and of the sea cow may well have been one
and the same. The teeth also are similar to those of early probos-
cidians.

Whales and seals have already gone back to the sea and by structural alterations committed themselves to sea life. It would be interesting to know what kind of creatures they were when they lived
on land. Unfortunately, fossils do not give us that information and
we have to go by deduction, and vague deduction at that. There
is no need, however, to feel frustrated. Life is a moving picture,
not a still photograph. Given the same conditions, what happened
once will happen again, and *is* happening again. If we look around
at the present time we see the same process taking place before
our eyes. Many animals we know well are now undoubtedly starting on the long and arduous journey to the sea. Almost any animal
that is becoming as much at home in the water as on land is a
possible candidate for sea life. There is no need to make a list—
the reader can do that for himself. But I will mention two mammals. Other vertebrates, including birds, have queued up for the
sea, as have some (though very few) insects, but this chapter deals
with mammals.

Some time ago, an intelligent member of the weasel tribe saw
a future in fish, and went after fish in streams and rivers. The otter
is now equally at home on land or in the water, but the odds are
it will not long remain like that. Already it frequently journeys to
the sea, and well out to sea.

This applies to the so-called Land Otters (*Lutra vulgaris*) and
the larger American (*Lutra canadensis*). A relative, similar in appearance but bearing (to its sorrow) much richer fur, is the Sea
Otter (*Enhydra lutris*). The Sea Otter has already *arrived* at the
sea, passed its tests, and said good-bye to the land. It breeds in the
sea now (though not in mid-ocean) and produces its young in
beds of kelp. It lives on sea food such as snails, crabs and clams.
After diving, if it brings up a clam observers say it often brings up
a flat stone as well. It then lies on its back, places the stone on its
chest and hammers with its paws to break the shell. It is a highly
intelligent animal, and a playful one too, but, as I say, it has a lovely

fur and that fur brought it practically to extinction. In America a single skin would fetch $1,700 forty years ago. Protection has enabled it to recover to a certain extent. If it makes good this recovery no animal will have been so close to the brink of extermination and got away with it.

Sea Otter diving

The polar bear might seem an unlikely candidate to those who only see it (and its attractive but bad-tempered baby) enclosed in grills in zoos where it only occasionally enters a small pool, but in natural conditions the polar bear from birth to death never sets foot on land. It only needs the melting of the ice floes to force it to a decision—to sea or not to sea. The Arctic ice is melting and breaking now. This may be just one of those periodical phases and of no immediate account, but the time will come when the earth flattens, figs grow in Greenland, and the polar bear finds himself between land and the deep blue sea. He is already as much at home in the water as on the ice and the odds are that he will take to the water. If he becomes another whale it is to be hoped that he retains some of the aggressive spirit that characterises him now. Right whales, had they possessed the cantankerousness of the polar bear, would not have been exterminated so quickly.

It is perhaps largely food that prompts the return of land animals to the sea. Even man himself is now casting envious eyes on the sea's plankton. Where, on land, could animals in their millions find sustenance enough to lay on the bulging layers of fat that whales and seals carry? Fishes do not store fat, but consider the flesh and weight a salmon puts on after a sojourn in the sea!

You may say there is another side to that picture; the sea is a terrible place where creature preys on creature and death is always round the corner. This is so, of course, but not more than on land. The expectation of life of an adult pollock, for instance, is almost certainly greater than that of an adult rabbit. A large percentage of the eggs and immature young of prolific species *have* to be destroyed or the sea would become solid with them, but the percentage is no larger than the percentage of, say, aphids that are killed by enemies on land. The young of eels have to travel for five years at sea before they reach the rivers of Europe. We do not know how many are taken, of course, but millions of them arrive at their destinations safe and sound. Similarly the mother eels who make the same long journey going the other way obviously manage to get to their breeding ground in the deep Atlantic in large numbers. The danger time for sea trout and salmon is not when in the sea but when in the streams and rivers of the land.

Of course, tunnies, dolphins and the rest have to live, but after their attacks the preyed-on shoals swim on without any apparent diminution in their numbers. The odds against an individual herring or mackerel or any member of a large shoal being taken by sea enemies must be great and since shoals, by their numbers, attract attention from marauders the chances of survival of less gregarious fish are probably greater.

Before they came up against firearms it was the same with the herds of game on land. In South Africa, lions, leopards, wild dogs and others took their toll, but never seriously reduced the numbers of the herds of the various antelopes. This is obvious, for if they had done so there would have been none left for the white men to exterminate at the end of the last century.

We are apt to view natural slaughter rather lopsidedly, and think it grim that a lion should pull down a beautiful antelope every so often. But the antelopes themselves think little of it. When one of their number is killed they do not, as a rule, stampede over the far horizon. They know that that particular lion will not trouble them for some little time to come, and, if they think at all, they probably think that when he does strike next some other member of the herd will suffer and not themselves, and this is a very sensible view.

Are we not the same? No lion stalks us now but there is another stalker amongst us, and yet when we hear of other people dying we merely say "poor old James," or whatever the name was, and resume the even tenor of our ways. We do not panic, though we know that sooner or later the stalker will get us—knowledge which in the case of the antelopes is hidden from them.

CHAPTER 5

FISHES: THE SHARKS

WHEN we think of life in the sea we automatically think of fishes. I gave the sea mammals first place only, because they followed on as a natural sequel to the land mammals previously mentioned. Really, fishes should have pride of place.

Whilst the creatures on land developed so did the fishes in the sea, though not in so spectacular a way. It is generally like this; it is the adventurers who change and not the stay-at-homes. Nevertheless, the fishes developed into a larger number of species than did any class of vertebrates on land, reached what must surely be perfection in underwater performance, and occupied all the seas, and every part of the sea—the beaches, the surface waters, the middle waters, and the depths.

It is not necessary to describe a fish. Everybody knows a fish, and those who require a detailed anatomical description must go elsewhere: they will get the information by consulting an encyclopædia.

Naturally, in view of their diversity, fishes have been divided into many sub-classes and a large number of orders, but there are only three main divisions, or classes, of which the first, Marsipobranchii (lampreys and hag-fishes), will not be allowed to detain us. They are without jaws, but possess sucking and rasping mouths after the fashion of leeches. They are a survival from the very first vertebrates (and so, in spite of their primitiveness, are in the same phylum as man) and belong more to the past than the present.

The other two classes are the Sharks (Selachii), and the Bony Fishes (Pisces).

It is the sharks with which we are going to deal in this chapter. Compared with sharks the bony fishes (or "true fishes" as they are sometimes called) are recent comers. The shark is a very old sea inhabitant. The first shark of which we have much knowledge was a fairly large fish called Cladoselache that swam the seas in the Devonian period, 300 million years ago. Its streamlined body gives evidence that it was fast, and therefore must have hunted active prey, and its large eyes and powerful teeth indicate that it was predatory. The large number of fossil remains found, and the number of different species, show that it was successful and that the seas at that time must have teemed with fish. These were bony fishes, but not the type we know today, the majority of them being lung fishes.

All seemed to be "Set Fair" for sharks; they were the rulers of the sea and they continued to multiply both in numbers and species. But as time went on sharks fell upon hard times. In the Permian and Triassic periods (200 million B.C.) the lung fishes began to disappear and most of the other bony fishes—the precursors of the modern sea fishes—were in brackish, shallow water. Starvation faced sharks, and those that survived had to change their ways: instead of swimming after fishes near the surface they had to move slowly along the sea bed, picking up what they could in the shape of shellfish and other insignificant fare. It was presumably during this difficult period that sharks developed the underneath jaws typical of so many species today, for this position of the jaws would make it easier for them to forage in the mud. Incidentally, this underneath mouth has given rise to the popular idea that sharks have to turn over to seize their prey. And when taking garbage from a ship or dealing with sinking corpses, they frequently do turn over and take the food from underneath, but when dealing with normal live prey they merely lift their heads when within striking distance and thus get the jaws in the right position. It is obvious that no shark could ever catch an active fish if it had to turn upside down before it could get hold of it.

However, better times (for sharks) were on the way. In the Cretacian period (about 100 million B.C.) the bony fishes from their

fresh-water fastnesses (where they had made considerable evolutionary strides and where they were becoming congested) came
sweeping back to the sea. The sea gave them more scope, they
multiplied and improved. These were the parents of all our mack-
erels, herrings, tunnies, and the rest; swift, active, and by no means
stupid, and if the shark was to take advantage of this new rich
source of food, he must again change his ways.

Some decades ago a picture called "The Man with the Muck-
Rake" (I forget the name of the artist) attracted much attention.
It depicts a man delving in a midden with a rake, looking down
and ignoring a glowing crown in the sky above. It is the kind of
allegorical picture the Victorians loved, but overlooks the unfortunate fact that a man must eat and that food is more likely to be obtained from the gross earth than from the rarefied atmosphere
above. The shark was now a bottom-feeder with an anatomy best
suited for lying on the mud, and in spite of the riches that had
come into the upper sea many of them remained at the bottom
and are muck-raking there today. These are the rays and skates.
Others, however, rose to the occasion and regained their former
activity and streamlined shapeliness. These (with exceptions) are
the sharks.

The chief difference between sharks and bony fishes is that the
sharks (and rays and skates belong to the same class as sharks)
possess a skeleton of cartilage, or gristle, while the bony fishes have
a brittle, limy skeleton.

There are other differences. One lies in the skin. If you pass your
hand over the skin of any shark it feels like emery paper. So does
the skin of the ray, except the electric rays that have a smooth skin.
The shark in short does not possess scales like a bony fish, but is
covered (as can be seen under a lens) with millions of little tooth-
like protuberances. And these protuberances are not only tooth-like,
they *are* teeth, for on careful examination they are found (like our
own teeth) to be made up of enamel-covered denture enclosing a
pulp cavity. Incidentally, these tiny protuberances on the shark's
skin ("dermal denticles" they are called) give us a hint of the origin
of our own teeth and of all teeth. The remote ancestral sharks had
only this rough skin, but by degrees the denticles on the skin at

the edges of the jaw became enlarged and developed into what we call normal teeth. These normal teeth (without the rough skin) were handed on to fishes and in due course to ourselves. Unfortunately our own teeth have not developed in the same way as the shark's; for some reason, ours, and those of our fellow-mammals, became embedded in the bone of the jaw. Furthermore, we were allowed only two sets, a miserly and inadequate provision. The shark's teeth are set in the gum and underneath each tooth lie a number of replacements ready at any time to fill the gap when the tooth above is lost. That sharks live to a very ripe old age is fairly certain, but however long they live they will always have a perfect set of teeth and if they lose a tooth a hundred times they will always get another. As to whether the setting of teeth in gum is adequate we need only study the shark's record; no creature that can bite through steel traces or tear dead whales to ribbons has any cause to be dissatisfied with its teeth. The perpetual replacement of teeth was handed on via the fishes for some time as is shown by a reptile, the crocodile, today, a creature whose teeth always "grow again." Why nature should have altered this arrangement with succeeding types is difficult to understand. With mammals, hair and nails and skin and bones grow again without limitation, but teeth, a very similar adjunct, do not.

The bony fishes have their gills set close together, covered by a single plate. The sharks have a number of separate slots, five to seven in number, on each side, that work to a certain extent independently.

The shark, unlike most of the bony fishes, has no air bladder.

The method of breeding of the bulk of the bony fishes is well known. The female lays millions of eggs which are afterwards fertilised in the water by milt from the male. Sharks are viviparous in most cases, bringing forth their young alive either directly, as with mammals, or from large eggs that hatch within the bodies of the mother. And the young are not to be counted in millions but by the half-dozen or so. A few species including the dog-fishes and some rays *do* lay eggs, large fertilised eggs enclosed in a hard leathery covering, that float about and in due course—sometimes as long as a year—hatch. Sexual union between the parents, therefore, always takes place, and the males possess what are called "claspers" for use in the sexual act.

Finally, though the flavour of fish may be of little biological importance, the flesh of sharks does not appeal to most human palates, while the bony fishes provide us with an attractive choice of appetising food. True, the pectoral fins of certain small rays and skates are cut off and command a ready sale, especially to fried-fish shops, and are nourishing, if insipid, food, but bearing in mind the vast number of species of sharks it is surprising what little use they are to the hungry dwellers on land. The famous shark's fin soup of the Chinese is delicious, but it does not owe its tastiness to any shark. The shark's fin (when used, which is not always!) merely provides a thinly glutinous base for what is added to it. Some sharks are eaten and liked by certain island tribes (so are crocodiles by certain tribes in Africa), but on the whole, the shark is not fit for human food and some species are purgative or even poisonous.

When referring to the bony fishes we always differentiate between the species. The fisherman does not describe his captures as just "fish." If he catches a trout he says so, and if it is a tunny, a mackerel, a cod or a conger eel he also mentions the fact. Yet to most people a shark is just a shark and that is the end of it. Even on the radio the other day a man who was supposed to be an expert described a certain fish as "as big as a shark." Since the various species of sharks are of any size from sixty feet to a few inches the comparison was not very informative. He might as well have described some land animal as being "as big as a mammal." In short, sharks vary in species and size, just as do the bony fishes. It does so happen that the bony fishes possess more species than sharks, but this presumably is just chance—unless there *is* some advantage in possessing a limy skeleton and bringing forth young in a haphazard way. A class gains nothing, however, by having more species than another class, and the sharks are now well abreast with their environment and enjoying great success. It would be interesting to see the state of affairs some 50 or so million years hence and learn the result of the Selachii v. Pisces battle that is raging now.

Enough has been said to show that a survey of the different

species of sharks is impossible here. We might, however, glance at a few of them.

The two largest sharks share most of the waters of the sea between them, the gigantic whale shark inhabiting the warm seas, and the basking shark the colder. Dangerous as they look, those who expect blood-curdling stories about these two monsters will be disappointed, for both live chiefly on plankton.

The whale shark grows to more than sixty feet and swims near the surface in the warmer waters of the Pacific, Atlantic, and Indian oceans and other seas also. Its huge mouth, wide as a frog's in relation to the rest of its body and capable, one would think, of swallowing an ox, is an intimidating sight, but, as we have said, this shark is a plankton-feeder, straining its food through sieves situated in the gills. Occasionally, however, that cavern-like aperture of a mouth may inadvertently take in larger things such as pieces of wood which perforce have to go into the stomach and remain there. The whale shark is a strong swimmer with a torpedo-like but stout body capable of chasing active prey. But, as I say, it prefers shrimps.

The basking shark reaches forty to fifty feet. It is a bluish sinister-looking shark whose only sign of its innocuousness (if one may take the whale shark as a guide) is a large, rather blunt mouth showing honestly in front of its face and not hiding viciously below as do the mouths of most of its relatives. It is found in temperate waters all over the world but particularly in the North Atlantic. It is migratory to a certain extent and its migrations may be connected with breeding, but little is really understood about the comings and goings of this fish. In Europe it is known, however, that it passes the coast of Ireland in the spring, heading northwards, on to the Scottish islands and on to the Shetlands, arriving off Norway in August.

It is obvious, therefore, that it can move to some effect when the migrating instinct propels it, but on the whole it prefers to live up to its name and spend its time basking in the sun, lying like a cat in almost any position from upside down to the normal, exposing now its back, now its side or belly to the sun—when there is any. Hence another of its names, the sun fish (not to be confused with certain bony fishes of the same name). It is indeed rather odd that such a lover of basking and sunning should keep to the cooler seas and cloudier atmospheres.

It is both solitary and gregarious. It may be encountered alone, in groups of two or three, or in shoals of about one hundred.

The basking shark's slothful love of ease very nearly led to its downfall. There is no easier creature to approach and harpoon from a small boat and it was hunted considerably off the coast of Massachusetts in America, and also to a lesser degree off Ireland, Scotland and Norway. What the hunters were after was its liver—nothing else; the rest of the massive carcass was useless. However, the liver of this shark, from the point of view of bulk only, is not to be despised. It is enormous. Perhaps this explains why this shark is so sluggish, for it is almost as if it suffered from an enlarged liver. A single liver will give two hundred gallons of oil (four hundred gallons in some cases) and this oil was used for a variety of purposes including tanning, and tempering steel.

But slaughter depleted the fisheries; the oil, too, became of less commercial value and the basking shark was allowed to go its own slow way.

Recently attention has been turned to them again and numbers are now killed off Wales and other countries. It is again the liver that is the prize. What the oil is used for now I do not know, but cod-liver oil is not always from a cod.

And there is no reason why it should be. The liver oil of almost any fish is probably just as good.[1] The cod happened to get on the market first, that is all. At one time cod-liver oil was used only by human beings (chiefly children) to counteract deficiency of sunshine vitamin D. The demand, therefore, was not very large, for no human being is likely to take more cod-liver oil than he needs, but nowadays it is used for domestic animals as well—dogs, cats, canaries, poultry, horses, cattle, pigs, and the rest. On almost every sack of poultry mash, for instance, you will see the magic words, "With C.L.O."

The basking shark, like the whale shark, lives on plankton. Since they can generally get right up to the animal, hunters in a small

[1] Provided it is a "white fish" such as cod, hake, haddock, or the flat fishes which store their oil in their livers, and not an oily fish such as a herring or a mackerel.

boat are in a position to select the best place for the insertion of the harpoon. The best place is where it will pierce into the internal organs and cause the most pain, thereby subduing the animal more quickly. In spite of this the normally sluggish creature puts up a tremendous fight. It dives to the bottom where it tries to rub off the harpoon. After that the fireworks start and it will draw out as much as twelve hundred feet of line and someone has always to be ready to cut the line to prevent disaster. It has sometimes taken as long as nine hours to subdue this shark and bring it to the side of the boat. The liver is cut out as the shark lies alongside and the carcass released to float away and rot.

Whales for their blubber, seals for their coats, sharks for their liver and the rest of the carcass discarded. Over the centuries man has wasted enough riches from the sea to have made his arable land twice as fertile as it is. These riches after a few days stink to high heaven, but if a thing is good enough to stink it is good enough to eat, after nature has put it through her careful processes. We throw back into the sea (which does not need it, having inexhaustible wealth of her own) stuff that *we* need, nasty stuff for a time but of far more intrinsic value than furs and oil.

Inoffensive though this shark is, hunting it is not a relaxation for the aged or infirm. When the battle is over and the shark drawn alongside, a single blow from its tail could stave in any rowing boat.

The great white shark or "man-eater" is a very different proposition from the two above. Almost as big as the basking shark (forty feet or more)[1] it has the ferret-like underneath mouth of the prey-hunters, and teeth three inches long. The word "white" must not be taken too literally; its back and sides are a bluish grey, only its belly is white.

Large as this shark is, there is fossil evidence that it was once much larger. In the Eocene period (50 million B.C.) it was a monster of over one hundred feet.

Its other name, "man-eater," must make us pause and digress about the so-called dangerous sharks. Do sharks eat men? When a man is lost overboard in shark-infested waters he is invariably presumed to have been "eaten by sharks." Is this supposition right?

[1] When the length of sharks, etc., is given it applies to large specimens. Fish vary much in size according to locality and other circumstances.

There has been a lot of argument. Many say that sharks are scared of a man swimming in the water and will not go near him. The human remains sometimes found in the stomachs of sharks are, they affirm, parts of corpses buried or drowned at sea and subsequently eaten by sharks. In adventure stories, of course, sharks are horrible things that tear to pieces any man they come across, with much reddening of the water—the latter being a favourite touch. Sailors, too, tell blood-curdling tales. I should like to relate some of these and one day I will, but what we want now are authentic records, and these are hard to come by. This is not the fault of the sharks but of the human beings. Human beings do not usually fall overboard in calm weather when sharks and qualified witnesses are present. Sailors are, alas, not infrequently washed overboard in rough seas and lost, but in those conditions nobody can tell whether they are attacked by sharks or not. Over the years there must have been many incidents, duly witnessed, that would have advanced our knowledge, but what normally happens to information of this kind is that the ship returns to port, the sailors tell the tale over and over again in the local pub, then they go back to sea, and in a couple of months the whole thing is forgotten. It was reported that many of the crew of a line-of-battleship wrecked off Cape Finisterre were "devoured" by sharks when trying to swim ashore, but whether this is really authentic I do not know.

The normal food of a shark is a fish and all animals are inclined to be dubious about food to which they are unaccustomed. A live man in the water, especially if he splashes about in his very non-fish-like way, may well be given a wide berth. Sharks are cautious creatures. I can testify to that, together with others who have tried to get one with a piece of pork jammed on to a filed-down butcher's hook. In my case the shark would display all the interest of a woman viewing a smart hat in a shop window, looking at it first from one side, then the other, but never making up its mind to take it—though it swallowed any garbage thrown overboard including, on one occasion, a tin can. Sharks, however, I admit, are often taken in this crude way.

In his book, *Men and Sharks*, Hans Hass relates how he photographed sharks (probably blue sharks) under water in his aqualung outfit, and none of them made any serious attempt at attack. On these occasions, however, the sea was full of dead or dying fish dynamited by the author (I make no comments on this proceeding!) and the sharks were fully occupied eating them.

Sharks avoid human beings simply because they are not accustomed to them, but when the novelty wears off it is a different matter, as the hundreds of bathing casualties show. Take that other so-called man-eater, the crocodile. Contrary to general opinion most crocodiles are pure fish-eaters—but they soon learn. An African potentate in Northern Rhodesia who had his kraal on the banks of a river got the bright idea of tying up criminals ("criminals" being people he disliked) and throwing them into the water. For a time they were merely drowned, but soon the local crocodiles became interested and learned to take a major part in these executions, much to the delight of the potentate. A shark could be educated in this way probably more quickly than a crocodile.

So do not think that you can go swimming from some ship in perfect safety when sharks are present. Quite possibly the majority would not attack you, but there might be one that had had dealings with human beings before and had lost its awe.

Another shark, I think, that would not hesitate to attack is a hungry shark. Sharks, of course, are always hungry, but I would put nothing past a really hungry shark—a starving one, that is.

I am not trying to concoct a sort of Highway Code about sharks, but never bathe in waters where sharks may be if you have cut yourself. The smell of blood makes any shark dangerous. When I was at Lobito Bay, West Africa, a native fisherman was dreadfully savaged by quite a small shark when wading from his boat to the shore. He had done this countless times before but on this occasion he had cut his thigh on some tackle in the boat and was bleeding.

Though there is not much authentic data about people being attacked by sharks in deep waters there is plenty of data about them being attacked when bathing near the shore. And this is rather strange because most dangerous species of sharks do not like water so shallow that they can hardly manœuvre in it. These attacks are by individual sharks that have become accustomed to shallow water and to the ungainly splashings of bathing humanity at seaside resorts and are attracted by the smell of flesh. Grills are therefore

erected at many bathing resorts in hot countries and notices posted
that if you do not bathe in the grill-enclosed areas then you bathe
at your own risk—though whose is the risk except your own, in
any case, I have never found out. I have bathed behind these grills
in Durban and other places and felt glad of them though it did give
one a feeling that one was one of a bunch of sea lions in a zoo
without the sea lion's swimming capability.

Statistics, again, are hard to come by. Casualties are recorded in
newspapers when they occur, but the length of life of one issue of
a daily newspaper is short. It is known, however, from records that
in ten years in Australia up to 1930 there were eighty attacks by
sharks on human beings, nearly half of which proved fatal.

As an aside, here is an account of the activities of a certain small-
sized specimen of the great white shark in Australia. I do not know
the year, but it could not have been recent, for there were two grills,
one for ladies and the other for gentlemen. Sir F. McCoy told the
story and J. R. Norman quotes it. "A specimen about 15 feet long
had been observed for several days swimming around the ladies'
baths, looking through the picket fence in such a disagreeable manner that the stationmaster had a strong hook and iron chain made
so as to keep the rope out of reach of his teeth, and this, being
baited with a large piece of pork made to look as much like a piece
of lady as possible, was swallowed greedily, and then, with the aid
of a crowd of helpers, the monster was got on shore. On opening
the stomach, amongst a load of partially digested objects, a large
Newfoundland dog was found, with his collar on, identifying him
as one lost the day before, no doubt swallowed while enjoying a
swim in the comparatively shallow water."

All sharks have a keen sense of smell, and I have already mentioned how the smell of blood attracts them. In warm waters a
harpooned whale will often draw crowds of sharks that behave like
hungry wolves when the dead or dying carcass comes to the surface,
continuing to tear at it even when they themselves have been mortally hurt by spears.

Many authorities laugh at the statement often made that a shark
bit off an arm or a leg "like a carrot." Examine, they say, the leg or
arm bones of a man or, as a substitute, the bone of a leg of mutton.
It would be impossible to bite through this. But when one sees

the jaws of, for instance, the great white shark in a museum (which is the best place to see them) the feat does not seem at all impossible. Furthermore, human arms and legs *have* been snapped off frequently by sharks. Even a large dog can crunch a mutton bone, while the hyena, no bigger than a labrador, can break with ease the thigh bone of an ox. Its jaws can also bend the barrel of a shotgun.

Sailors dread sharks and their hatred of them is notorious. When one is captured sailors will often inflict every torture on it they can think of. A favourite trick is to take its eyes out and set it free again so that its end will be as prolonged and miserable as possible. There must be some reason for such hatred, passed on over centuries. It can hardly be merely because sharks eat corpses. Many fishes and every crab or lobster does that—the sea would be an unclean place if corpses were not devoured. Incidentally, this occasional fiendish treatment of live sharks by sailors is probably a waste of time except in so far as it relieves their own feelings. So far as the evidence goes sharks do not seem to feel pain, or, at any rate, to be much affected by it. I have told how they will continue to attack the dead body of a whale even when mortally wounded. There was a case, too, when a captured shark was cut open, its guts removed, and the body thrown back to the sea—only to be caught a second time on a hook baited with its own entrails.

The great white shark is found in all the warm seas of the world. It eats what it can get, and that is usually fish, though one specimen was caught that had a whole sea lion in its stomach, and we have already mentioned the Newfoundland dog. It is not in the habit of following ships for the sake of the garbage.

This shark is another of the large sea animals that have been credited with the feat of swallowing the prophet Jonah. Linnaeus selects it in preference to the sperm whale. And it certainly could have swallowed Jonah whole and, had the prophet survived the digestive juices, could also have ejected him in a more or less intact condition. For many sharks from time to time have a clear-out of their stomachs, even turning that organ inside out so that it projects several feet from their mouths.

The great white shark usually has a litter of about nine young. Little is known about the breeding habits of sharks in general or

what attention, if any, they give to their offspring after birth. The
ancients credited them with great maternal devotion, insisting that
when danger threatened, the mother swallowed all her little ones
and kept them safely inside her until the crisis had passed. That is
as may be, but sharks and their young are rarely found together
which indicates that the young lose no time in getting away from
the devotion of their mother. In any case, they get no nutriment
from her.

Every passenger on a slow-going ship gets a thrill the first time
the famous triangular fin of a shark is seen moving alongside. The
sharks that accompany ships are nearly always "blue sharks," lithe,
beautiful creatures that sailors call the "Wolves of the Sea." They
have the most insatiable appetite of any creature and eat anything
living or dead. They roam everywhere and spend most of their time
on the surface. Since they will follow ships for days or weeks, it
seems possible that they never sleep.

The largest species of the blue sharks is the great blue shark,
which may grow to twenty feet. It is universal in warm seas and
sometimes found in colder seas, though it never grows large in cold
seas.

There are superstitions galore about sharks in general, and about
the blue sharks it is said that when there is going to be a death
on board they gather round. I remember one old sailor who almost
convinced me at the time that he was speaking the truth (and he
had certainly convinced himself). He gave the name of the ship
and the names of the officers. I wish I could relate the story as he
told it to me with all the trimmings, but I cannot. Briefly, the
mate had gone down with fever and dysentery. He was very ill for
some days, then seemed to take a turn for the better. Everyone was
pleased until that afternoon, when the fins of a large concourse of
sharks were observed around the ship. That night the mate took a
turn for the worse and died in the middle hours of the morning.
He was buried at sea next day and my friend doubted if the "bas-
tards" got him. They'd put an extra lot of lead on him and he went
down "nice and quick" while the "devils" above were fighting
amongst themselves.

Probably this superstition originated from the early days of sailing ships, when typhoid, smallpox and other epidemics often broke out amongst the crowded emigrants of all nationalities. Human corpses might then be consigned to the sea almost every day, attracting an ever-growing following of sharks. It is not a far step from this to attribute a special sense to sharks, and a little embroidery will do the rest. Anyway, there are old salts going about today who firmly believe that sharks know if a man on board is going to die. Nowadays of course most ships travel too fast ever to be accompanied by any sharks, so the legend is bound to die a natural death. A pity in a way; this is now a too-prosaic world and a legend or two does nobody any harm.

We cannot make a list of the "dangerous" sharks, there are too many of them, but it ought to be mentioned that, in the opinion of many, the palm for savagery goes to the Tiger Shark (*Galeocerdo tigrinus*). It is particularly active off the West Indies and in the Indian Ocean and is dreaded in those areas more than any other shark. It is a large shark but built on the slender comely lines of

Monk or Angel Fish (above), *Wobbegong* (below)

the blue sharks. It is accused by natives of a large number of murders.

The tiger shark, however, is interesting from another angle than that of mere ferocity. Usually any shark brought up in the fishermen's catch is regarded as an unmitigated nuisance. It tears the nets, eats the baits, hooks and all, and bites the fishermen themselves even if it does not knock them half-stunned into the sea with its tail. The one and only problem is to get rid of it. But the tiger shark is now deliberately being hunted for the sake of its skin. Previously the skin was useless, but recently, skilled workers, using special apparatuses, can do the very delicate work of removing the dermal denticles without tearing or injuring the skin. The result, after treatment, is a good-looking leather, and a harder-wearing leather than any other we know. We need a hard-wearing leather. For some reason, which no one seems able to explain, the hides we use today have not the durable qualities of the hides produced, say, thirty years ago, in spite of (or is it because of?) modern methods of tanning. The tiger shark is also said to provide very superior cod-liver oil, and possibly does, but they were bound to say that. Having caught and skinned their tiger shark they naturally wish to use the liver and a little boosting does no harm. Durable leather is now also being made from the skins of other species of sharks and the time may come when shark leather is in general use, and when we are indebted to the sea for yet another everyday commodity.

The shark is essentially a warm-water fish. It is more numerous in warm seas, more active, and grows larger. The basking shark and a few others inhabit temperate seas, but only one large shark lives right up in the cold and ice of the Arctic. This is the Greenland shark, a massive, thick-set creature, but fast enough when after prey. It spends a good deal of its time in the bottom reaches of the sea, but also hunts actively on the surface. Like most sharks it eats anything it can get, and probably fish form its staple diet, but it appreciates a change of fare and is a perfect scourge amongst the seals. Above all things, however, it likes food that is stinking and rotten. It gathers in large numbers to eat the offal that goes into the sea from the salmon and other canneries in Alaska and New England

Hammer-Head Shark

—and this refuse is far from fresh. The Greenland shark is the supreme scavenger.

A whole reindeer was found in the belly of one specimen, so had this shark ever inhabited the Mediterranean regions we should have had yet another Jonah-swallowing suspect.

There are many accounts of packs of Greenland sharks attacking right whales in the old days when those whales still existed. It is now ruled that such attacks must have been made on dead whales, not live ones—though I can see no reason why a pack of these always-hungry, formidably-jawed, twenty-foot-long sharks should *not* attack a defenceless animal only three times larger than themselves. If they have never done so then it indicates a great lack of enterprise.

Greenlanders, too, have their stories about being attacked in their *kayaks*—and certainly they are very scared if a Greenland shark appears when they are at sea in their frail boats. But many authorities shake their heads. The Greenland shark, they say, does not attack human beings. And I hope they are right, but much as I respect the opinion of those who should know I would not trust myself in the water with any shark accustomed to snapping up swimming seals five hundred pounds in weight. With the best will in the world it might make a mistake.

The flesh of this shark used to be used extensively by Greenlanders and Icelanders both for themselves and (especially) their dogs. Strangely enough, the meat is poisonous when fresh but wholesome for both men and dogs when putrid or preserved by drying. The process of putrifaction evidently destroys some toxic property that is present in the flesh when fresh. Incidentally so-called "tainted" meat (such as a well-hung pheasant) is more di-gestible than fresh meat, and the Eskimos always try to keep their seal meat until it is "high" before eating it.

Not all sharks are of the conventional shark shape. Take the wobbegong, or carpet shark, found off the coasts of Australia, China and Japan. This is a squat fish with a round head edged by a mous-tache formed by tassels of skin that look like bits of seaweed and give it quite a pantomime appearance. Its body is patterned like a carpet and it spends most of its time half-buried in the mud. The broken carpet pattern makes the body indistinguishable from its surroundings.

Another peculiar-looking fish is the hammer-head shark. There is no other fish in the sea like it. Its head is shaped in the form of a double hammer by two large horizontal lobes on each side that are three feet across. At the ends of these lobes (which in a hammer proper would be the striking surfaces) are set the shark's little eyes, one on each side. It is hard to see any advantage in this arrange-ment, but the hammer-head shark flourishes, grows to a large size, and is found in every sea.

Amongst examples (there are many) of unusual shapes in sharks, the angel fishes should be mentioned. They are so called not from

Blue Shark with Pilot Fish

any spiritual qualities but from the resemblance of the curve of
their pectoral fins to the curve of the wings of angels as conventionally depicted by most artists. They are also called monk fish
because there is a resemblance to the cowl of a monk on their heads.
They are small, flat, sea-bed livers. They are half-sharks and half-
rays, and are of great antiquity. As fossils show, they have remained
unchanged since Jurassic and Cretaceous times, 100 million years
ago.

Sharks have been known to take to the habit of piloting ships
into harbours. The famous Sydney shark will be remembered, and
"Pelorus Jack" of New Zealand. Why they should do this is a mystery. It is also a mystery why pilot fishes should accompany sharks
just as if they were a specially hired escort. For almost any large
shark will have its attendant pilot fishes, keeping their fixed positions like outriders, as the shark pursues its majestic course. Anon,
one of them will dart away to investigate some object in the water
and if this is edible will return and communicate the information
and lead the shark to it. If the object is not edible the pilot fish
will merely take up its position again. This, at any rate, is what
used to be firmly believed and it may be right, though it is hard
to understand why a shark, with its keen sense of smell, should need
assistance in locating food.

Passing on information to a shark, however, is not so far-fetched
as it might seem, for similar associations between the small and the
large occur on land. The honey-bird of Africa, for instance, a bird
no bigger than a sparrow, will deliberately lead men or honey-
badgers to a bees' nest, twittering loudly to them as it does so, while
they respond with calls or grunts to show that they are following.
The association here is obviously of benefit to both parties: the
bird cannot break into the nest itself but the men or badgers can,
and are bound to leave enough honey and grubs spilt around to sat-
isfy a small bird.

But if we concede that the pilot fish leads the shark to food how
do we explain the fact that pilot fishes will also accompany ships?
A ship will not turn aside and follow them for food, yet they have
been known to escort a sailing ship for three months, only leaving
it as it entered harbour. And it was not for the sake of the garbage
thrown overboard, for they never touched it.

Sharks are generally hungry. They were born hungry. Why do they not eat the pilot fishes? Because, it has been suggested, the pilot fishes are too agile for them. But would pilot fishes journey with sharks for months if they were in constant danger of being snapped up? Because they are too small for the shark's consideration, say others. They are nearly a foot long, and I doubt if any fish is too small for a hungry shark's consideration. Nor are pilot fishes poisonous to eat. Those who have tried them say they are very good—like mackerel, which is not surprising, for they are of the mackerel tribe.

Pilot fishes do not depend entirely on the leavings of the shark, for some of them caught with sharks have been found to contain small fishes in their stomachs.

It is probably right to sum up by saying that pilot fishes accompany sharks (1) for protection, and (2) for the food the shark leaves. But this does not explain everything. When a shark is caught and is being hauled up from the water, the attendant pilot fishes, instead of streaking off to safety, often show evident signs of distress, swimming round and round the body as it is drawn up almost as if they would like to rescue it and pull it back. No, there is some mysterious association between the great sharks and these small fish.

The pilot fish is a gorgeous creature when freshly taken from the sea—emerald green with broad ultra-marine stripes, and a blue tail of which both extremities are tipped with white. It is found in most warm seas. It is only large sharks that they accompany.

CHAPTER 6

FISHES: THE RAYS

Those who regard evolution in the sea or out as a gradual but inevitable process of improvement should consider the Selachii and compare the rays with the sharks. They would conclude that the sluggish, lumbering, flattened, often tailless rays must have preceded the active, streamlined sharks, whereas, as we have already seen, the reverse was the case: the rays evolved from the streamlined sharks and only put in an appearance towards the middle of the Mesozoic era (100 million B.C.).

Sharks and rays are both Selachii, but this class has been divided into two orders: Pleurotremata (sharks) and Hypotremata (rays), the latter being sharks that have fitted themselves to a life on the ocean bed.

Vastly different as most sharks are from most rays the dividing line is not at all clearly marked. As colours laid down side by side on an easel may merge together at the inside, so do the orders of the sharks and rays, and there we may find sharks that look like rays, and rays that look more shark-like than the ray-like sharks. How then, of these so similar creatures, can one say that this is a shark and that a ray?

It is not difficult; for the ray in the flattening-out process it adopted in order to be able to lie like a piece of carpet on the sea sand or ooze had to do something about its gills. In the shark the gill slits are at the sides, but when the ray became flat it had no "sides" to speak of, so that the gills had to go somewhere else. The rays, therefore, evolved underneath gills and these were reduced

in size. It is by these underneath gills that one may distinguish the
Hypotremata from the Pleurotremata.

The switching of the gills to underneath may sound a simple
solution of the ray's difficulty, but it entailed one drawback. A shark
breathes by taking in water through its mouth and discharging it
through its gill slits. But the ray, half-buried as it usually is in mud,
would breathe in through its mouth not clean sea water, but a mixture of water, mud, and sand like the dregs of coffee. Such a mixture
would inevitably damage the extremely intricate arrangements in
the gill-slit exits. So the ray had to make another adaptation and
do what no shark can do and take in water from the spiracle or
nostril on the top of its head instead of through the mouth. This it
does only when half-buried in mud; when swimming in clean water
it breathes through the mouth in the ordinary way.

The first thing that strikes one about many of the rays is their
bat-like appearance, and swimming in the murky depths near the
sea bottom they must appear very like huge slow-flying bats. No
wonder the larger ones strike terror into the hearts of divers as they
pass over them, their great wings shutting off the light filtering from
above. Unlike other fishes the squat rays do not use their tails for
swimming but move their elongated side fins up and down like
birds. That is why the sides of rays are always referred to as wings
and not fins. Incidentally, one of the reasons why many rays do not
use their tails for swimming is because they have not got any, or
at least only possess a rear appendage that resembles a bit of cord
and looks like the tail on the end of the flat kites children sometimes fly.

As with the sharks, rays have a bewildering number of species
varying from the minute to some of the biggest monsters in the
sea, though the average casual fisherman is quite unaware of the
fact. There exist for him merely small rays and large rays and
medium-sized rays, and if he catches a saw fish he may not class it
as a ray at all. The rays are as diversified in their habits as in their
size and appearance. Nearly all of them, however, bring forth their
young alive and copulate like the sharks.

Sea fish are known to most of us solely through the medium of
the fishmonger's slab, and, indeed, these slabs are little museums

in themselves and can be studied with interest and profit even by those who do not wish to buy, though such must not expect sympathetic collaboration from the fishmonger himself. These slabs, however, are not very instructive to the student of sharks and rays for he will find there no whole specimens of either. He will, however, almost invariably find portions of one species of ray, which will be labelled with the name SKATE. I may add that this was not always the case; when I was a boy, fishmongers in the north sold it always as RAY (and may do now), and it is as ray that I remember the flavour and texture of this fish.

Actually, there are many species of skates, and all are of the genus Raja. Some species are large and some small. It is the small or young skates that are best for food. Only the wings are used, the rest of the body is discarded, generally as soon as the fish is caught. It is only comparatively recently that skate has been accepted as an article of diet. Gastronomically the skate began life as a poor man's fish, scorned by all who could afford even the cheapest of the other fish, but helped by this cheapness (it was almost given away), it made rapid headway and now occupies quite a respectable position on the slab. Indeed, on the Continent, specially prepared, it has become a dish for epicures. So this once-despised fish that when caught was always thrown back into the sea is now becoming scarce through over-fishing. To me, skate is an insipid dish, but I know those who prefer it even to turbot and salmon.

It is not necessary here to study the several species that come under the heading of "Skate." There is a species found off North America from Nova Scotia to Florida; another off California, another in the Japan Seas, while the "Common Skate" (R. *batis*) chiefly inhabits the seas of West Europe.

The skate is a very exceptional ray in that it likes cold water. Rays in general love warmth and become fewer in colder waters. Skates pursue their sedentary lives in all the temperate seas, but are most numerous in the Northern Hemisphere, where many of them live close to the Arctic Circle.

The skate spends most of its time lying on the sea bottom, and its camouflaged back gives no indication that it is there. It waits for something edible. When this edible thing—a fish perhaps—passes by, it is not met with the flashing fangs and swift pursuit to which it is accustomed but finds itself covered with a sort of rug: the skate has thrown itself on it and covered it with its smothering body and wings. All the skate has to do now is to feel for the prey underneath and eat it.

When in the mood the skate can swim well, flitting by with rhythmic movements of its wings over the sea bed.

Where skates are known to be, there are anglers who deliberately fish for them with rod and line. It is one of the most unexciting forms of sport I know, though there is a trick in it, if you can call it a trick to haul in with all your might at the first pull on the bait. If you do this, provided the skate is not a large one, it comes up like a heavy stone. If you give it a second's margin it flops down into the mud, forms a vacuum underneath and almost needs a steel cable to dislodge it.

Most of the rays, as I have said, produce their young alive. The skate is an exception. After mating, the female gives birth not to little skates but to eggs—dark-coloured eggs about the size of duck eggs. These are flat, brown and leathery and have two prongs at each end. These prongs attach themselves to weed at the bottom and in due course the weed grows and surrounds them in a safe embrace. Before the weed has grown over them, however, they are apt in rough weather to break loose from their moorings and be washed up on to the beach where any holiday-maker may come across them. These stranded skate eggs have various local names such as "Mermaid's Purses" or "Sailor's Purses" and, indeed, they have rather the appearance of small rectangular purses.

In spite of the cold seas in which the majority lie, the eggs hatch in periods that vary from six months to a year and a half.

As a complete contrast to the skate let us go now to that large and dangerous ray, the saw fish. In almost every port in the hot portions of the globe, tourists will find saw fish blades for sale at curio shops. They may also be found in private houses, and I know of no uglier ornaments.

The saw fish wears its teeth set in sockets on both sides of a six-foot-long, flattened, blunt-nosed blade that projects in front. It has another much smaller set in its mouth. Now many people spin yarns about saw fishes attacking boats and thrusting their blades through the sides. Similar tales have appeared in print. These people are confusing the sword fish (a bony fish) with the saw fish. There is no excuse: both fishes are aptly termed, one carries a saw and the other a sword. The sword fish can and at times does, pierce the sides of a boat with its sword, but if the saw fish tried to do the same thing with its blunt-tipped saw it would merely break its nose off.

What use then is the saw to its owner? It might be thought that this very awkward weapon sticking out six or more feet in front of the fish would be a mere encumbrance. Actually, it is one of the most useful tools a fish ever possessed. It has four uses that we know of, and may have others.

(1) It is used as a trowel to grub along the sea bottom and dislodge edible creatures that lie there.

(2) An ordinary fish pursuing a shoal of smaller fishes must chase and catch them individually; the saw fish gets amongst a shoal and flays its saw right, left, and centre, cutting dozens of fish into pieces that can then be devoured at leisure.

(3) The saw fish can attack a fish of equal size without fear of retaliation; a few slashes with the six-foot saw will incapacitate another fish at long range and after that small portions can be slashed off for consumption, for this fish does not possess a large mouth.

(4) Attacked by a larger fish, the saw fish can use its blade equally well for defence at a distance.

On the other hand, the sword of the sword fish can only pierce, and if it pierces either prey or foe the sword fish has to extricate its nose before it can do anything more. (It can at times, however,

Saw Fish

when amongst shoals of smaller fish, use it as a flail after the manner of the saw fish.)

Most rays are mere flat discs, broader than they are long. The body of the saw fish is more like that of a shark. The fins are on the large side giving the creature a somewhat frilly appearance, but it has the lines and speed of a shark. It is that hall-mark of the rays, the underneath gills, that stamp it as Hypotremata. It attains a length of from twenty to thirty feet of which one-third is saw. It lives in almost all the warm seas and occasionally enters the tidal waters of such rivers as the Mississippi and the Zambesi. It brings forth its young (about twenty-five at a time) alive.

Lastly, the inevitable question: is the saw fish dangerous to man? It certainly is not a natural "man-eater," if that is what is meant, and it has no interest in men, but it is dangerous enough. Those who deliberately hunt it, as some natives do, run risks, for some time or other they have to come within reaching distance of the saw. And there is on record an account of an Indian enjoying a quiet bathe in the sea who was cut clean in half by one slash from a saw fish. Indeed, some Indians fear it as they fear no other fish, including sharks, and there must be some basis for a reputation like that. Hooked from a boat, as one might expect, it will tow the boat and occupants mile after mile; a free trip marred only by the thought of the end when the occupants come into closer contact with the fish—and its saw!

The flesh of the saw fish is coarse and tasteless, but the animal is a valuable catch purely for its fins which go to China to make shark's fin soup. And, of course, the saw to the curio shops.

Of all the creatures in the sea the rays are usually regarded as the most uninteresting, and to fill his hall a lecturer must label his subject "Sharks" and not "Rays." Yet rays provide far more diversity than sharks and when you meet them there is rarely a dull moment. Forced by hunger in the first place they have branched out more, and given rise to species possessing remarkable adaptations. One ray, as we have just seen, developed a double-edged saw. What next? The next is a ray that went in for electricity and used it long before Leyden had thought of his jar.

The electric organs of the electric rays (as they are called) are exceedingly complicated and only a genius in the field of electricity could fully understand them. I am no electrician myself so the description I give of them will be brief and written in hope that the reader will understand the mechanism better than I do. These electric organs are a complicated wet battery. There are about four hundred fifty special tubes in each of the organs supplying the positive and negative currents, all separated from each other by special insulating tissue. There are many electric plates, and the wet medium (corresponding to the acid solution in a man-made wet battery) is a clear jelly. There are special nerves going to every plate which come from a main nerve which itself is connected to a separate section of the brain that deals solely with electricity.

Nature, in fact, rather surprisingly, has shown herself here (and in other electric fishes) to be a skilled and inventive electrician. I say surprisingly because with most other animals she has given no hint that she knew anything much about the subject, and in the inanimate world her best-known efforts—thunderstorms—are rather crude affairs.

True, whenever we or other animals use our muscles there is a minute discharge of electricity, but it is a far cry from that to the intricate apparatus of the electric rays.

Skipping more detailed examination of the insides of the batteries of these rays it may be said simply that externally they consist of two nodes, one positive and the other negative, which have to be connected before a discharge takes place. A powerful shock is then given of a frequency of as high as one hundred fifty per second. This kills small creatures and is quite enough to knock a man flat on the ground. A faint shock can even be felt in the arms via the net when fishers are hauling in a catch that includes one or more electric rays.

The strength of the shock varies greatly according to the size and condition of the fish and the number of electric plates it brings into play. Should its battery be fully charged and should it bring all its electric plates into play (an unusual combination) the effect on a man who trod on one would be spectacular.

The electric organs are used deliberately by the ray both for defence and attack. Apart from this an electrical discharge is given spontaneously when, for instance, the ray is trodden on as it lies buried in sand.

Why nature should have fitted out these particular rays with electricity is difficult to understand, for normally they lie on the sea bottom like other rays, waiting for what comes—small fishes, crabs, etc., and never use their intricate apparatus. When they *do* use it they electrocute their prey and then consume the unconscious body. Frequent use weakens the battery and it needs a period of rest to charge up again. Any attackers, seeing what appears to be an ordinary ray of the right size for a meal, get a most unpleasant surprise when they come to close quarters.

Most electric rays are small: one of three to four feet long (and the same across) would be a very large specimen. They are of the pancake shape, practically circular except for the small shark-like tail which clearly shows their origin, and which is used for swimming. Like most of the rays except the skates (Raiidae) they bring forth their young alive. They are more common than is thought and are to be met with in numbers off the Atlantic coast of the United States, off California, and in the Mediterranean. They are no great gift from the sea to us for their flesh is watery and tasteless. It is, however, fairly innocuous to the consumer, which cannot be said for the flesh of all the rays.

Not very long ago a shock from an electric ray was held to be a certain cure for gout. It is no longer used as a remedy for that complaint, possibly because live electric rays well charged with electricity are hard to come by, or because patients prefer the gout.

Do not confuse the sting rays with the electric rays. Both are unpleasant but their unpleasantness takes different forms.

Sting rays (there are some thirty species) have the flat, almost triangular body of the common skate and are broader than they are long. This is not taking into account the rat-like tail, which is the root of all the trouble. Some prefer to compare the tail to a whip, and these fishes are called whip-tailed rays. These tails are fitted with barbs, one to three in number and varying in length from three to twelve inches, and these barbs when lashed against some unfortunate creature inject poison.

It is a curious fact that the ancients have often been ahead of recent generations in natural history knowledge, in spite of the fact that they did not maintain an army of fully-qualified scientists. For instance, two thousand years ago they said that ants stored seeds, an idea laughed at by moderns until about a hundred years ago, when it was found that ants *did* store seeds, and on a large scale. Similarly the ancients said that sting rays injected poison from their tails, and this statement gave quiet amusement to authorities until about fifty years ago. There was no poison, they had said, in the barbs on the tails of these rays. How could there be? There was no mechanism as with wasps, spiders or snakes for injecting a poisonous fluid. True, a man slashed by a tail suffered pain and swelling, sometimes leading to gangrene or tetanus, but this was caused purely by the dirt and slime in which the ray lived and which coated the barbs on its tail; which was reasonable enough and which certainly accounted for any cases of gangrene or tetanus. Yet Pliny had said, "Nothing is more terrible than the sting that arms the tail of Trygon.[1] It is as strong as iron, yet possesses venomous properties." Pliny, when he let himself go, was an exaggerator if ever there was one, but on other occasions he was an acute and reliable observer. This was one of those occasions, for it was found (though it took two thousand years to do so) that the barbs of the sting ray *do* inject poison. A small groove containing a powerful venom runs along the edge of each barb. It is easily broken open and then the poison flows into any wound made. We do not know the chemical formula of this poison any more than we know that of bee and wasp venom.

The effect of the sting on a human being varies according to the size of the fish, and, no doubt, the conditions and age of the human being. In extreme cases it can cause death with convulsions. It can also cause the loss of movement in a limb for a long time, insensibility for a short period, rarely more than a few minutes, and in all cases where the poison escapes from the ruptured spine into the wound, severe pain.

The broken barb, I am sorry to say, is quickly replaced: there is always another one ready to grow out and take its place.

The sting ray uses its tail solely as a whip, and not for swimming, lashing out vigorously at prey and enemies alike. There is a small species in Australia that often lies buried in the sand on beaches

[1] The sting ray. The name Trygon now denotes the genus, and the family is called Trygonidæ.

used by bathers. When trodden on, its tail whips out from under
the sand, punctures the bather's foot and inflicts agonising pain.

Throughout their known history human beings have always (provided they were at the right end) been fond of the whip as an in-
strument of correction. They have from time to time invented many
kinds of whips; corded whips, whips with metal attachments, whips
with thongs, etc., so it would be surprising if the whip of the sting
ray should have been overlooked. In Ceylon it was cut off, treated
with oil to make it pliant, and used for flogging criminals. Used
thus, of course, the barbs contained no poison, but the punish-
ment nevertheless was so severe that this corrective practice had to
be abolished by law.

Sting rays flourish in all the warmer seas. From our viewpoint
(disregarding the pain they sometimes inflict on us) they have no
merit. The flesh of the electric ray is flabby, insipid and watery.
That of the sting ray is this and more: it is positively nauseating.

Inhabiting the warm seas is another family rather like sting rays,
but larger. These are the eagle rays. There are only about a dozen
species of which the best known are:

(1) The American eagle ray, which swims along all the warm
Atlantic coasts of North and South America.

(2) The Californian bat fish, which is numerous in the mud flats
off California, and

(3) The common eagle ray, which passes most of its time in the
Mediterranean and the Atlantic seas adjoining.

Many specimens have a length of well over twelve feet and a
weight of eight hundred pounds. Their tail is similar to that of the
sting rays but the poisonous spines are nearer the base. The spines
are therefore not so "handy" for offence—but they are handy enough
and the one ambition of the fisher who catches one is usually to
get rid of it at the earliest possible moment. In a boat it gives
savage grunts and barks and lashes its tail, apparently with the deliberate intention of striking any occupant. A Dr. Cole relates, "On
the morning of 12th July, 1910, while handling a large specimen it
suddenly threw its body against me and drove its poisoned sting
into my leg above the knee for more than two inches, striking the
bone and producing instantly a pain more horrible than I had

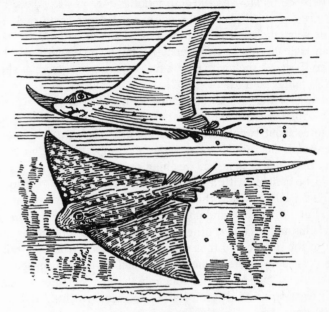

Spotted Eagle Ray

thought possible that man could suffer." After tearing the barb out Dr. Cole injected an antiseptic solution which relieved the pain. This is interesting, for no antiseptic solution I know of will relieve the pain of a bee or wasp or scorpion's sting. There are authentic cases of permanent disablement, of the loss of a limb, and even of death from eagle rays.

Though flat and clumsy-looking, eagle rays can swim well. They can also fly! At least, they can break the water and skim over the surface for a short distance, hitting the water again with a resounding flop.

Most of them feed on molluscs such as clams and oysters. Oysters, particularly, have so many enemies—worms, borers, squids, octopuses and the rest—that one wonders that man has ever a chance to eat any himself, but let one or two eagle rays get amongst planted-out beds of oysters or clams, and that is the end of the bed.

Not one will remain and the destruction is done in a few days. This will give one some idea of the power of the jaws of an eagle ray. A three-pound clam would, with us, need a large hammer to crack, yet an eagle ray, not by hammer action but by mere pressure of bite can crush such a shell with ease, a feat requiring a pressure of about one thousand pounds.

Largest of the rays is the manta, known also as the devil fish, the vampire ray and the sea bat. It is as broad as it is long and measures about twenty feet across the disc. Its big gaping mouth gives it a rather stupid and yet alarming appearance. One would think that such a monster, weighing over three thousand pounds would be content with a life on the sea bottom like so many of its smaller relatives, and this is the life it undoubtedly lived in the past. But a three-thousand-pound body needs a lot of nourishment, so for this and no doubt other reasons the manta brought its huge bulk to the surface and took to swimming about. It had no tail to help it (it only possesses one of the small rat-like appendages) so had to use its "wings" beating solemnly through the water like some squat legendary bird. It soon learned to swim gracefully and easily in the upper regions of the sea, a feat accomplished by no other tailless creature of that size. Sometimes when pearl divers are at work a manta comes along and moves silently by like a great cloud blocking out the light over them. This takes several years from the lives of the divers for they have been told that the devil fish will descend, enfold them with its wings, draw them to its mouth and swallow them. Actually, the manta rarely swallows anything much larger than a sardine or a herring: it lives chiefly on plankton-feeding crustaceans as do the basking and whale sharks. Strange how so many of the greatest monsters of the sea live on the smallest type of food.

It is remarkable that the manta should be able to swim well and to live in the upper regions of the sea, but what is more remarkable is that it should be able to jump several feet right out of the water like an eagle ray. When the eagle ray jumps it makes, on return to its native element, a loud splash; when the three-thousand-pound manta returns, belly first, it makes a noise that can be heard for miles. An observer who was in the vicinity of a jumping manta com-

Manta Ray

pared the noise to that of the discharge of a cannon "reverberating away."

The wonder is the manta does not injure itself in these sportive jumps. I remember in Chefoo, China, a man trying to emulate a girl who dived gracefully from an alarmingly high sea-wall. She had told him that from that height he must not dive too steeply, but must commence his dive in an almost horizontal position. The weight of his head, she said, would bring his body to the correct angle when it struck the water later. He obeyed instructions and commenced his dive in a horizontal position, which position, for some reason, his body maintained until it met the sea, the diver by this time having assumed the sprawling attitude of a chicken thrown over a fence. He had to be helped out of the water and later had to see a doctor, and he weighed nothing like three thousand pounds, nor measured twenty feet across the belly. But the manta seems to like it, and that is the chief thing.

There are many legends about this giant ray and one is that it will grasp an anchor between its fins, pull it up, and make off with it—including the boat attached behind. So any yachtsman, having dropped anchor, may suddenly find himself proceeding to sea at speed.

The manta produces only one young in a year. This infant is about five feet across and weighs about twenty pounds.

A smaller edition of the manta is the mobula or lesser devil fish, and it is small only by comparison with its bulky relative. Norman tells us that when landed, the manta emits harsh grunting barks, whereas the mobula produces a melodious bell-like sound.

CHAPTER 7

FISHES: THE BONY FISHES

WHEN the bony fishes, after a period of development in rivers, lakes or brackish estuaries, swept back into the sea, those older inhabitants, the sharks, gave them an eager reception. A new era had begun for them, an era of plenty. But a new era had also begun for the bony fishes and in spite of shark persecution they multiplied and spread over the oceans and penetrated to every depth until they became the real possessors of the sea. The sea possesses countless other forms of life, crowding the bottom and jostling microscopically close to the surface, but the possessors of the open spaces of the sea are the bony fishes.

After their arrival these fishes branched out and evolved into ever-improving forms, helped in the first place by the sharks, for improvement comes only from difficulties or persecution. Later they got even more persecution from their own kind, which assisted the development both of the persecutors and the persecuted still further.

Many bony fishes are not unlike many of the sharks. A sword fish, for instance, has a general resemblance to a saw fish, and a plaice lying on the sea bottom could easily be mistaken for a skate. And the bony fishes differ greatly amongst themselves; a turbot is very unlike a conger eel, while some of the fishes inhabiting the depths are like nothing anywhere else. We have already mentioned the fundamental differences between the Pisces and the Selachii, but will go a little more fully into some of them.

The chief difference, of course, is the limy skeleton of the first.

The shark gets on very well with a skeleton of gristle, but the limy skeleton is a later model of fish chassis, and later models are generally considered to be improvements. In the beginning the bony fish (who probably originated from some shark-like creature) tried several devices, one of which was to enclose himself in a suit of armour, thus wearing his skeleton outside instead of in. This device evidently did not succeed, for it was discarded.

Another difference is that in most cases sharks bring forth their young alive and in small numbers whereas the vast majority of the sea bony fishes lay multitudes of eggs which are only fertilised after they have emerged. This may not seem a very advanced thing to do, but it succeeds and is now a part of the economy of the sea, a piece of the pattern. The shark takes all from the sea and gives nothing back except its carcass; the bony fishes give back much and help in the sea currency exchange.

The bony fishes shed their eggs direct into the sea, and having done that in the places most suitable to receive them, forget all about them. These rich globules, full of nutriment, then lie floating (or on the bottom in the case of the herring and its tribe) amongst a mass of hungry creatures all on the lookout for anything edible. Nature fashioned each of these eggs with as much care as she gives to the human egg, but what are their chances of survival? To get over this the female, instead of laying a few dozen, lays large numbers. Out of a large number one or two ought to get through—though there is no real certainty about it. Of the so-called food fishes the ling takes the least chance of being left childless by laying 160 million eggs at a time, though the sunfish easily beats this by laying 300 million. At the other end of the scale comes the herring with a mere 30,000. The herring, therefore, lays only one egg to every 10,000 of a fish like the sunfish, yet it outnumbers every other species. So much for one fish; when you multiply the number of eggs by the millions of cod, for instance, or herring in a shoal you get figures that are truly fantastic, as are the numbers of larvæ hatched from them.

To manufacture all these eggs, each furnished with a yolk that the larva will carry about and feed on for many days, in view of the fact that only a few of them, if any, will come to maturity, seems wasteful. The way of the sharks seems better. But, as I say, it is all part of the economy of the sea. When their yolk sac is consumed the young join the minute eaters of the copepods and diatoms, and so take their place in that essential chain that distributes the plant life of the sea.

Most of the bony fishes possess air bladders, sometimes called gas-bladders, or swim-bladders. The last two terms are better than the first for the bladder rarely contains a mixture of nitrogen and oxygen in the proportions of air: the bladder of fresh-water fishes usually contains a large percentage of nitrogen, that of sea fishes contains more oxygen—indeed, the oxygen is sometimes almost un-diluted. Incidentally the gas-bladder can be used as a source of oxygen to the muscles of the fish, on lines rather similar to the lung of the lung fish.

Swim-bladder is the best name of all, for the chief function of the bladder is to enable the fish to "float" at certain depths. By regulating the gas pressure he can move about on a horizontal plane at any reasonable depth without any draw upwards or downwards. Land dwellers (and sharks) can never know such ease. Even when asleep on a feather bed there is, with us, always that pull of gravity that must unconsciously be counteracted.

That amateur dissector, the housewife, when she guts and cleans a fish, knows the appearance of the swim-bladder well—the silvery sac of varying shapes near the backbone.

The housewife, however, does not dissect the head—she lets the cat do that. If she did probe into the skull she would find another difference between bony fishes and sharks. In a cavity on each side of the hard skull are two chambers, each containing a small stone. These are the ear stones, or "otoliths," and these chambers and stones constitute the ears of the bony fish. These ears are very dif-ferent from the ears of land dwellers, and quite how they operate is not known. They are probably connected in some way with a nervous region called the lateral line, which we are going to mention in a moment. In themselves they cannot be nearly as efficient as our own ears, but (like ours) they are certainly used for detecting vibrations. Sharks do not possess such stones.

If the housewife *can* find these stones (it is not easy and requires a strong knife) she might be interested to examine them, for they will tell her the age of the fish she bought. Not that from her point of view the age matters, for fish, unlike chickens, do not become tough when they grow older, but the interesting fact is that they *do* tell the age. The stones, which are flattish and of different shapes according to the species, are marked with concentric rings, each ring denoting a year's growth (rather after the manner of scales). In old fish, interpretation becomes somewhat difficult except to an expert, but up to about ten years it is easy.

How long does a fish live? I have mentioned the theory held by some that because a fish feels little of the pull of gravity and never really stops growing it can live for ever—well, I think we can forget that theory, chiefly because it can never be proved. Few fishes have a chance of living long, and the bulk of them never have a chance of living at all. Also, though the strain of gravity may not affect them when at rest and undisturbed, they do not remain undisturbed for long. An enemy appears and they have to flee in all directions, upwards, downwards, sideways. This must tend to cancel out the absence of strain due to buoyancy. Carp, however, live very sedately in ponds and lakes and seem to have no age limit. Specimens, and not very large ones, over a hundred years old have been taken. Herrings twenty years old and plaice thirty years old are not uncommon.

The lateral line stretches from the head down the flanks to the tail. It is a narrow, nervous region, very sensitive to vibrations. It enables fish to swim in shoals without colliding and probably warns them of the approach of enemies. It enables even blind fishes to avoid obstacles.

The lateral lines of no two species are quite the same, and here a knowledge of anatomy helps the housewife. No one can fail to distinguish a cod from a haddock, but with fillets of these fish it is a different matter—or would be but for the lateral line. The haddock has a dark line, the cod a white line. Fishmongers are very worthy men, but not all of them can resist taking advantage of a greenhorn (a trait not confined to fishmongers) and smoked fillets of cod are cheaper than smoked fillets of haddock. Housewives would probably never fall for the trick, but their menfolk evidently do—at least, I have had a fishmonger try to sell me a fillet of smoked cod as smoked haddock, *and* to argue the point. To do him justice he had no haddock in his shop and was only doing his best.

Incidentally, another distinguishing mark of the haddock is a kind of black thumbmark on each side towards the front. This is sometimes called the "Mark of Christ," and sometimes "Saint Peter's Thumbmark." One legend is that this mark was imprinted when Christ fed the multitude with five small fishes, another that it was set when Saint Peter pulled this fish out of the lake of Genessaret—in spite of the fact that no haddocks could possibly have existed there.

Mention of senses calls for a few words on the senses of smell and sight. The smell sense of fishes is very acute, more acute apparently than we realised until quite recently when salmon fishers are being advised to de-odorise their waders and their tackle before going near the water. Smells in water ought, of course, to be more perceptible than smells in air, though not for such long distances. On land we can only smell gases, and those gases have to be dissolved in fluid before they can register. The fluid is the saline moisture in our nostrils supplied by small sacks—another reminder that we have not escaped sea conditions yet. Smells come to fish already dissolved and they detect them instantly.

Izaak Walton realised this two hundred years ago, but until recently his directions in *The Compleat Angler* for giving bouquets to baits have been held as interesting but rather ridiculous. So sensitive are fish to smells that apparently a mere strong smell containing no toxic properties can kill. On the other hand, they will approach and eat rotting carcasses, and carps feed and thrive on sewage. This need occasion no surprise; the dog, an animal endowed with a sense of smell that is probably a hundred times more acute than ours, is attracted by the odour of putrifaction. So are we in our cruder way when we eat a well-hung pheasant or partridge, for smell is the chief contributor to taste.

Fishes smell with their noses, and in the bony fishes this is the only function of the nose. It does not, as with land vertebrates, serve also as a breathing passage. In addition to the sense of smell fishes have also an acute taste sense. The taste buds, however, are not, as with us, situated in the mouth but all over the body surface and even on the fins, so that most fishes can detect food just as well with their tails as with their front parts, and as well with their backs as with their bellies. This is possible, of course, because they live in water and do not have to confine their taste buds to an area specially watered by saliva. And it has yet another sense that we have largely lost—though not so much as we might wish. This sense enables the fish, through the surface nerves of its body, to appreciate changes of salinity and faint traces of irritant chemicals. So keen is this sense that a chemical in the water—of such dilution that a naked human bather would never perceive it—will often agonise a fish to death. Our own skin is now impervious to such substances, but this was not always the case. In our fish and amphibian stage, we were equally susceptible, and are so now on those surfaces of our body that are kept permanently moistened—the eyes, nose and mouth. Smells have nothing to do with it; the smell of a cooking onion is quite pleasant, but peeling and cutting an onion, though it has no effect on the rest of our body, will hurt the surface nerves of our eyes and nose and cause smarting and watering. Ammonia will do the same thing, and so, as a more extreme instance, will tear-gas. Imagine then our discomfort if, like the fish, these slight traces affected our whole bodies.

As for the sense of sight, the most remarkable thing about the eyes of a fish is that they are basically so like our own. Were no other data available a biologist might be excused if he hazarded a guess that, purely on account of their eyes, fishes and men must be related in some way. Actually, the sight of a fish is rather like that of an elderly man, the lenses are inelastic and cannot change focus, so since the fish cannot put on spectacles it only sees an object clearly when it is at a certain distance. A trout, for instance, waiting for food drifting down the current, stays in a fixed position. It sees a blur approaching and waits until that blur gets to the right focal length when it can sum it up for what it is or is not, and take or leave it. In a still pond it has to swim to the correct distance.

Probably it is unable to distinguish colours, though artificial-fly fishers will never believe it.

The fish is cold-blooded, its internal temperature being the same as the temperature of the water. An apparent exception is the tunny. This active fish, weighing up to a thousand pounds, which can swim at nearly fifty knots—and would thus be able to overtake and pass a torpedo—has an internal temperature a few degrees above that of the water. To call it warm-blooded, however, would be going too far, and it may be that it is only after terrific exertion that its temperature rises.

The introduction of central-heating into mammals was a remarkable thing and one of the chief reasons for their rise to dominance. Using their food not only as food, but as fuel to keep burning an interior stove, mammals can go about their business unaffected, within limits, by outside temperatures. The only drawback is that the warm-blooded animal *must* keep his stove going: once his stove goes out he goes out with it. Mammals and birds, therefore, cannot endure long periods of fasting that mean nothing to fishes and insects. All that happens to a fish when it fasts is that it ceases to grow. This may seem to be contradicted by the long fasts of—for instance—seals at mating time, but they are not really fasting, they are feeding on the huge quantities of fish and squids they swallowed over a period of nine months, a good portion of which, like some provident housewife, they stored up in the larder in the shape of blubber against the time it would be needed.

The disadvantage to a fish of the lack of an internal stove is its susceptibility to changes of temperature. Rod and line fishermen have always some excuse to make for catching nothing, and a favourite one is the temperature of the water. Sceptical friends sometimes smile, but they need not. A slight fall in temperature means nothing to us because we remain at the same heat inside. A fish's temperature falls with that of the water. When our internal temperature does change even by a few degrees we call it "running a temperature" and go to bed and do not eat. A fish, too, feels ill and does not eat when the water temperature alters. Indeed, a sudden change of a few degrees is often enough to kill them. A woman I know, commiserating with her goldfish in cold weather, gave them some nice warm water and seemed not only surprised but indignant when they all died.

As an instance of the effect of falls of temperature on large shoals of fish let me recount the story of the tile-fish (*Lopholatilus chamaeleonticeps*), a species unknown until 1879. In that year a schooner set out from Gloucester, Massachusetts, trawling for cod. Very few cod were taken, but suddenly the nets became full of large handsome fish of a kind the skipper had never seen before. He threw away most of them and was sorry later, for the new fish had apparently come to stay, and became a public favourite in the Boston and New York fish markets. It was, moreover, abundant. Then, one day in 1882, the boats went out and did not catch a single tile-fish, although the trip before they had come back heavily laden with them. Some days later steamers arriving at New York, Boston, and Philadelphia reported that they had passed through large quantities of dead tile-fish. Later it was ascertained that the dead fish covered an area of about five thousand square miles and rough estimates gave their number at 100 million.

And that seemed to be the end of the tile-fish, for not a single specimen was taken for thirty-three years. Then, in 1915, they suddenly reappeared in their former numbers, and have been abundant ever since (a recent catch was 1,238,500 pounds).

Various reasons were given at the time for this mysterious disappearance: disease, a subterranean upheaval, poisonous gas from the sea bed, etc., all very unlikely in view of the fact that fishes of other species remained unharmed. It is now believed that they were killed by cold, for the tile-fish is a dweller in warm currents and more susceptible to cold than cod and other food fishes. Exceptionally cold conditions had prevailed in the year of its disappearance, and even the lower, warmer water had been invaded by ice.

When I said that the bony fishes had now occupied all the seas I did not mean that they were plentiful everywhere. Fishes in the sea need certain conditions if they are to multiply in the spectacular way some of them do. One of the requisites for supporting huge populations is shallow water, and by shallow I do not mean water one can paddle in but anything up to about two or three hundred fathoms with access to underwater plateaux, sometimes called "banks." There are many such areas, notably along the west and east coast of North America (where lie the famous Newfoundland Banks) and along the continental shelf that extends from the north of Norway to the sea off Portugal, including the areas round the British Isles and off the Netherlands and Denmark. There is also the raised ocean bed, bordered by abysses of deep Atlantic that stretches from the Faroes to Iceland and on to Greenland. Farther east are the shallow seas off North-West Russia and round Japan. These areas are the richest fishing grounds in the world.

Similar conditions obtain in other parts of the world and doubtless these contain similar concentrations of fish, but knowledge of fish populations and species comes to us not from scientific expeditions (though these do gain much information), but from intensive and regular fishing by fleets of trawlers, and such fleets are only economical and possible in areas within reasonable distance of large human populations and markets able to dispose of everything that is caught. On land, naturalists can make expeditions and map out the numbers and species and migrations of animal life in every part, but they cannot do that in the sea. For a comprehensive knowledge of fish life in the sea we have to fall back on commerce. Therefore, our only really accurate knowledge is confined to the northern seas. The Antarctic seas hold just as much plankton and any shallow seas there may support a rich accumulation of fishes, but the distance of big markets will prevent commercial exploration for some time to come, especially since the Antarctic does not possess large areas of shallow seas.

The warm seas, too, are the home of many fish, from the ubiquitous flying fish to the monsters that pursue them. Never in my life have I seen so many fish as when sailing in a tramp steamer near the St. Vincent Islands. They covered the mirror-like sea like simmering scum, large fishes and small, up to the horizon on every side. What they were doing on the surface like that I do not know, but the skipper said he had seen it before in that locality. Hardly any sea birds were feeding on them; probably they had already taken so many that they were completely surfeited, an unusual condition in a gull. Most of the species of warm-water fishes are known, but we are still very ignorant about their movements.

The main requisite for supporting large concentrations of fish is, of course, food. We have gone into that already when talking about plankton, but an interesting fact is that several rich fields of plankton are partly dependent on the land. We land dwellers, one way and another, take a lot from the sea, not the least of which is about half our supply of meat (if one can class the flesh of sea animals as such). And this meat supply is raised for us by the sea—we have only to go and get it. But the business is not *entirely* one-sided.

Some seas where fishes team owe not a little to the land. Take the
North Sea, at one time the most crowded fish pond in the world.
On all sides rivers drain into it: the Rhine, Weser, Elbe, Thames,
Trent, Ouse, Tyne and others. In fact, a goodly portion of the most
fertile areas in Europe is drained into the North Sea. Every year
the cream of the rich top soil of the land pours into it—and this
has been going on for thousands of years. This river silt penetrates
far and is well charged with nitrates and phosphates and other
minerals necessary to plant life. Wave action on the shallow bottom, currents and sunlight keep this basic plant food in circulation,
and the diatoms, copepods and the rest thrive on it. Indeed, so
abundant are these small organisms that it is estimated that if a
square mile of sea water in the North Sea were evaporated their
skeletons would leave a residue of many tons of lime. (This, of
course, is one of the sources of the tremendous quantity of lime
required for the skeletons of the bony fishes, not to mention the
shells of crustaceans and molluscs.)

As we mentioned in an earlier chapter, much of the North Sea
was once dry land. Ages ago the sea swept up and covered it. Since
then the rivers have been doing their best to fill it in. Its average
depth is only sixty fathoms. In many places the spire of a village
church would show above the surface. Should this large area once
again be raised above the sea we should have land of almost incredible fertility. We should also, no doubt, have bitter wars for its
possession, so perhaps it is better where it is, for man has not yet
got to the stage of fighting over fish, though at times he has come
very close to it.

Similarly, fertile silt is deposited in the areas round North America by rivers and currents, and disintegrating glaciers bring detritus
from Greenland.

Moving currents improve conditions for fish. Waves and tides,
though they may look impressive, are not real movements of water.
Waves do not travel, though they may appear to do so, they merely
ruffle the surface while the water stays where it is like tea in a tea-
cup, and tides only tip the cup slightly from side to side. A current
is a different matter. It brings new food and also acts as a revivifying
agent. The Gulf Stream (I oversimplify here for the sake of brevity) moves along the east coast of America from the Gulf of Mexico,

and a subsidiary branch moves up the west coast of Britain and through the English Channel into the North Sea. The Baltic, being an inland sea, receives practically no currents and therefore contains no large fish congregations. Herrings, it is true, are fairly plentiful, but the herring (an exception in many ways) does not object to brackish water.

But fishes, of course, do not just stay put on their rich feeding grounds like sheep in a field. They have their migrations, and individuals of certain species can become rovers. These migrations are chiefly concerned with spawning and food, while temperature and salinity also play a part. We are learning something about these migrations by means of marking fish with metal tabs, but for obvious reasons it is difficult to learn very much. We know, however, that herrings do not make the long migrations into the deep Atlantic that they were once supposed to do, and we know that plaice, once supposed to be almost static will occasionally travel two hundred miles. Many of these migrations are, broadly speaking, on a fixed pattern. The shoals go to suitable spawning grounds, and after that follow the food—the herring, for instance, following the plankton, and the cod and a host of others following the herring, though other sources of food may deter or deflect them. Meanwhile, other fish may have made a migration to where the herring has laid its eggs and be having a royal time guzzling them up from the sea bed (for herrings, almost alone of sea fishes, lay eggs heavier than water), while others are engaged in eating the cods' eggs that for a short period float like masses of jelly on the sea surface. A large number of tunnies spawn in the Mediterranean areas and then, ravenous after their long fast and sexual exertion, go in search of food, but they move on a fixed plan like a knowledgeable fisherman who knows exactly where fish are to be found at certain times of the year. A portion of these tunnies (we do not know much about the rest) with unfailing regularity make for the North Sea in search of herring, arriving there in autumn, and never at any other time, and eagerly awaited by rod and line fishers at Scarborough who, fully armed with expensive gear, hope to get the thrill of their lives and great renown by hooking a big one. Yet again, migrations may be due to water temperature or salinity, to which fish are very sensitive.

In short, studying fish migrations is like probing into wheels within wheels.

We have spoken of the herring and will now go a little more fully into its movements, especially its movements round the British Isles, for here we have a good illustration of how easy it is to be led astray about migrations, and jump to wrong conclusions. As is the case off the east coast of Canada and the United States, the herring appears off the coast of Britain at certain times and constitutes one of the most important of all the fisheries.

The shoals first appear (apparently from nowhere) off the north-west of Scotland in May, and the fishing port of Stornoway in the Hebrides does its best to reap a harvest in a limited time from these truly vast congregations. They go, and in June are swarming round the Orkneys and Shetlands. In July they arrive off the north-east coast of Scotland, where the Aberdeen fishing vessels are eagerly awaiting them. Here their numbers decline after a few weeks, but in July and August they arrive in full force off north-east England, and the herring boats of Scarborough and Grimsby go into action. In October they appear farther south and now it is the turn of the fishing fleets of Yarmouth and Lowestoft, aided by the Scottish fisher girls who gut the fish and travel south by stages, in readiness at the various ports for the arrival of the herrings. By January the herrings are (or rather used to be) off Plymouth.

Now what was more natural than to conclude that all this was an annual herring migration? The times of disappearance and of arrival at the various places fitted in. Everything fitted in. So this is how it used to be explained:

After their tour round the British Isles and off Norway the herrings went into retreat, far out into the deep waters of the North Atlantic and the Arctic. When spring came, following the growth stages of the plankton as it "ripened," they moved towards land and in May a vast concourse of millions upon millions found themselves off the west coast of Scotland. Some then went southward along the Irish coast, but the bulk went northward and rounded the British Isles via the Orkneys and Shetlands, where they met other shoals pouring south from the Arctic. South they all went, first to Aberdeen, then on to Scarborough, Grimsby, Yarmouth and Lowestoft;

the other and slower-travelling shoals getting to Plymouth by January via the west route.

This was the general idea some time ago. There was difference of opinion regarding a few minor details, but everyone thought that a migration of herrings took place round the British Isles and to Norway. And several think so today. Only last summer a fishmonger said to me: "The herrings are grand this year. Those you got last time came from Aberdeen, they'll be coming from Grimsby soon, they're on their way there."

And herrings undoubtedly *were* on their way to Grimsby—but not from Aberdeen.

For we know now that herrings travel very little. They live offshore in comparatively deep water for most of the year and only gather together once a year for a few weeks when they come inshore to spawn. Why they do so at different yet definite times in such a circumscribed area as the British Isles we do not know. Food, water temperature, etc., probably contribute their stimulating effect, but these conditions vary annually much more than do the movements of the herrings.

When do herrings spawn? The herring, the best-known of all fishes, is also the most mysterious. It is always disobeying the rules. Normally fish spawn at a certain season once a year; the herring spawns twice, in spring and autumn. At least so it seems. There are now differing views on the matter. Some say that herrings do spawn twice a year, others that there are two kinds that differ not only in their season of spawning, but also physically, the one kind being spring spawning, the other autumn spawning. Others maintain that herrings only spawn once every eighteen months, so that a herring that spawns in the spring will not spawn again until the autumn of the following year, while one that spawns in autumn will not spawn again until the spring of the year after that. It is generally accepted now, however, that an individual herring spawns once a year like the other fishes.

The movements of herrings and other fish seem fixed; an annual affair, as seasonal as the seasons themselves and one that has been going on for ages. But at times we get startling and often unpleasant reminders that nothing is permanent. Over the centuries the herring shoals congregated off Plymouth in the month of January.

Plymouth was considered a fixed place of call in their annual pilgrimage. Then, in the early '30s, they came no more to Plymouth and have never been there since. There are theories about it, of course, but no one really knows why. Sardines (which are small pilchards) in the Bay of Biscay used to come inshore every year and the Breton fishermen depended on them for their living. But one year they failed to put in an appearance and the Breton fishers were in dire straits. The finest, largest, best-flavoured plaice in the world used to be found off the Faroes. They, too, disappeared and hardly one is to be caught there now. This, however, is not a very good example, for their disappearance may have been due to over-fishing. It is a puzzle, too, why certain species will congregate in one place and avoid other neighbouring places that have exactly similar conditions.

Fish may also appear or reappear in places where they are not expected. In 1929 there was a large increase in the cod population in the Far North, especially round Bear Island. It occurred at a fortunate time for the trawlermen, for previously the numbers of cod had been diminishing seriously. What had happened was that in that year there had been a general warming up of the northern waters—and this state of affairs still goes on. It does not follow, however, that it is permanent. Similar temperature changes have occurred before, but have rarely lasted more than a few years. It may be that the ice age is still on the wane and that we are on the way to one of the warm periods that sooner or later have always followed ice ages. We shall never know of course; the process is too gradual. Cold and warm conditions will still alternate, and a thousand years will bring little real change.

Whatever the purpose of their journey, a school of fish on its way is a wonderful sight. Moving swiftly through the water, following the leaders, every movement is executed with the precision and unison of a trained ballet corps. Their flanks flash simultaneously as all turn together. Packed closely, they move upwards or downwards or change direction without ever colliding. Whence comes the directive signal that causes these co-ordinated movements we do not know.

The bony fishes of the northern fishing grounds, the so-called

food fishes, are of two kinds—round fish and flat fish. The description is sufficient so I need hardly say that round fishes include such
types as the herring, cod, haddock, hake; while the flat fishes include plaice, sole, halibut, turbot, brill. In the beginning of their
evolutionary history the flat fish were not flat fish but round fish
like the others. This was the case also with the rays, and both rays
and flat fish flattened themselves in order to be able to lie unob-
trusively, half-covered with sand or mud, on the sea bottom where
their presence would not be noticed by prospective prey, or by
enemies.

So both these types, once round, became flat, and superficially
very similar to each other, but there was a great difference in the
way they acquired their shapes. They had to be squashed somehow;
but the rays were squashed from the top and the flat fish from the
side. The rays, therefore, though flat, are symmetrical, and the head,
except for its squatness, is as it was before, with the eyes in the
same position. If a fish is squashed from the sides it will have to
lie on one of those sides when on the sea bottom, and the side
it lies on will then become, so to speak, its belly. This has happened with the flat fish.

But a more difficult problem this side-flattening involved was in
respect of the eyes. If the flat fish proposed, as it did, to lie on its
side, then one eye would be underneath in the mud, a wasted member. So nature, that wonderful adapter, moved the bottom eye
round to join the other on the other side of the head.

We know this from a study of the fry in the early periods. Many
forms of life in their initial development, both in the egg and afterwards, re-enact for us, in a series of quick changes, the evolutionary
history of their race. The tadpole, for instance, gives us a broad
idea of the change-over of fishes to amphibians. The fry of the flat
fish show us in six weeks what occurred during millions of years.

We will take the plaice as an example—though any flat fish will
do.

On emerging from the egg the young plaice, almost microscopic
and quite transparent except for its little black pinpoints of eyes,
is an ordinary round fish with eyes in the position of those of round
fishes and (after it has righted itself and ceased to lie upside down)
swimming like a round fish. But after a month a strange thing hap-

pens; the left eye begins to move. Meanwhile the body slowly flattens sideways and the fry, a surface swimmer so far, begins to sink slowly towards the bottom.

The left eye is still on the move and by six weeks has reached the top of the head. A week later it has gone right round and has almost reached the right eye.

By now the young plaice has sunk to the bottom and is lying on what was its left side but which from now on will be its underpart—the white side—and the two eyes are close together on what is now the top of the head. (In the sole the two eyes practically touch each other.)

At one time it was thought that the left eye travelled part of the way towards the right by, so to speak, a subterranean route, actually disappearing into the skull and emerging later at a different point, like a mole breaking surface. This impression was given by the fact that the eye in certain species travels under a fin and so seems to disappear inside the skull.

With a ray, flattened from the top and bottom, the original underpart remains the underpart. With a flat fish, flattened from the sides, either side, surely, could become the underside. Yet, with plaice, soles, dabs, flounders, halibuts it is always the left side that is selected, and were it not for species like turbot and brill, we should probably still be puzzling our brains for the reason. There is, of course, no reason. It has to be one side or the other, and with turbot and brill the reverse flattening process takes place. In these fish it is the right eye that travels towards the left and the right side on which they lie.

We have finished with the plaice as an example of the sideways flattening of the flat fishes, but will tie up the threads. Having settled on the bottom, the lower portion of the plaice remains white, but the upper grows dark and takes on the familiar red spots. On muddy bottoms, as in the Icelandic seas, the plaice becomes very dark, almost black in fact, but on sandy bottoms such as those in the neighbourhood of the Faroes it is merely honey-coloured. When transferred from dark to light beds, or vice versa, the plaice changes its colour quickly. Judging from experiments the change is controlled by the eyes, for a light-coloured plaice laid with its head on light-coloured sand and the rest of its body on a dark background will remain light coloured, but a dark plaice in the same position will turn to a light colour all over, while the reverse takes place when the head of the fish is placed on a dark background with the rest of its body on a light background.

Except when migrating, the adult flat fish stay at the bottom and get their food there. An exception is the halibut. This large fish (one arrived at Billingsgate market weighing a third of a ton and needing six porters to carry it)[1] frequently leaves the sea bed to chase fish near the surface. It will be remembered that the very large rays such as the manta do the same thing. Perhaps as these fishes in the course of evolution grew larger, their diet at the bottom became insufficient for their frames and they had to supplement it by hunting fish in the upper levels. One of the biggest problems for all really large animals is food. The elephant in Africa as we have said before, has to be ever on the move to find enough to live on, though when it was the size of a donkey (as it once was) a small field would have sufficed. And perhaps food shortage was a major contributing cause of the extinction of the mastodons and the other outsize mammals.

Though the herring, cod and haddock remain fairly abundant, the flat fish are growing scarcer. Even on the wonderful Newfoundland Banks halibuts have decreased, though not, as yet, cod. Man is the culprit. It is naturally on those fishing grounds that have been longest fished that the fish are most scarce and show the red light to any newer fishing grounds that care to notice it. The North Sea, for instance, once the richest fishing ground in the world, has had its bed scraped incessantly for two hundred years by fleets of British, French, Norwegian, Dutch, Danish, German and Swedish vessels. It is the flat fish that suffer most, for the small ones are taken together with the mature fish. These small ones are quite useless for food, but owing to their disc-like shapes cannot slip through the meshes of the nets like the young round fishes. Furthermore, the trawlers haul up or crush large quantities of the small crustaceans, worms and molluscs on which these fish live, as well as the plants that harbour and hide them. So the Dogger Bank, which

---

[1] This was many years ago. Intensive fishing has reduced the size of this fish, since few have now a chance of growing large.

once sustained vast shoals of large and heavy plaice, now only yields
a few migrants to the trawlermen.

It used to be argued that so large are the fish populations of the
sea that no efforts of man could possibly reduce them. Thomas Huxley held this opinion. And we know the saying, "There are as good
fish in the sea as ever came out." This is not so. There are not as
good fish and not as large fish, and not as many fish as there used
to be—particularly flat fish. Certainly at the fishmonger's we see
plenty of plaice and soles, while halibut and turbot are rarely ab-
sent. But to get these fish the fleets must now go to far-off fishing
grounds.

It is a depressing thought that in a hundred years or so an acute
shortage of fish may add its quota to the general problem of feeding increasing human populations with diminishing food supplies.

There are twenty thousand species of bony fishes, so that any
attempt to sort them out and give descriptions in the space of one
chapter would be futile. I brought up the food fishes because they
are the best-known, because we have some knowledge of their movements and way of life, and because they are of importance to us in
our efforts to get as big a share as we can of the food the sea provides. This is not to say that the other bony fishes are little-known
or are not of importance as food. As a matter of fact, the average
man is probably better acquainted with the class Pisces than he is
with almost any other animal class.

And even to the average man there is more in fishes than mere
food. There is beauty; tourists have been known to miss their lunch
through being preoccupied with gazing at the brilliantly-coloured
fishes moving about coral reefs. Others admire the grace of movement of the fishes they see, their anatomical perfection for underwater progress, and their effortless speed (do not forget that some
fishes can move through the water faster than a racehorse can move
on land). Nor do those who keep aquariums have designs on the
flesh of the inmates. And finally comes that army of men and
women who find in fishes their greatest joy in life. Every holiday,
anglers in their thousands go off to their selected fishing grounds.
I know a man over sixty in Tunbridge Wells with a hard job in a
warehouse who gets up at five o'clock every Sunday to go to the

sea, where he takes a boat and spends the whole day fishing, whatever the weather. I know a naval officer, too, who invariably spends
all his leaves doing the same thing. Spinning plugs for bass, trailing
spoons for mackerel, etc., big-game fishing off New Zealand and
elsewhere, fly-fishing for salmon and sea trout, it is all an irresistible
lure to the addicts. Non-fishers, it is true, view this addiction with
bewilderment. Indeed, mankind seems to be divided into two
classes; those who fish, and the remainder who wonder what on
earth the fishers see in it.

However, the appeal that fishes may or may not make to human
beings is not a matter of great importance here.

The coelacanth is a bony fish and since it has figured somewhat
prominently in the news recently I will make some remarks about
it before closing this chapter.

It attracted attention by being described as a "living fossil." The
term is self-contradictory but everyone knows what is meant. Popular nomenclature generally manages to hit the mark without waste
of words. A living fossil has been described as "a relic of a fossil
group thought to be extinct discovered in a living state." Here extinction, or supposed extinction, is brought up as if it were a necessary preliminary to being a fossil. Actually, the discovery of a living
specimen of a species thought to have been extinct for many millions of years makes that specimen no more a living fossil than
many animals that are quite familiar to us and that we have known
for a long time. In a distant age a group of primitive animals was
isolated in Australia and in the absence of enemies such as later
evolved in the areas they had come from, escaped extinction and
change. So here in the flesh are antiquated forms that must be
sought for elsewhere in the rock strata of ancient periods. Amongst
them are the platypus of Australia and the tuatara of New Zealand,
and these are none-the-less living fossils because they have never
been classed as extinct.

All of us in a way are living fossils, for none of us was specially
created recently, but I suppose a fair description of a living fossil
is a creature that has not changed its specific form over a long period. How long this period has to be is a matter of opinion—it has
to be a good long time, that is all one can say. Into this category

come a large number of creatures including the king crab, the monk fish, the turtle, and, in our own homes unfortunately, the silverfish.

Vegetation, of course, also has its living fossils. Any botanist could name several. A plant that has come to notice recently is a conifer (*Metasequoia*, a relative of the gigantic Californian Redwood). This tree was found growing in the interior of China not many years ago, though scientific records had put it down as being extinct since the Cretaceous period. An American expedition went out to China about 1947 and took seeds which they distributed with great generosity. It is now growing well in several parts of the world, and growing, it is reported, far more quickly than most modern trees. It has been suggested it may be the answer to our wood problem. It has been given the attractive name of the Dawn Redwood.

Naturally the finding of a living specimen of a species thought to have been extinct for millions of years is of interest to scientists. They have already pored over fossil remains and made anatomical deductions. A living specimen gives them a check-up on the deductions they have made as well as supplying them with additional information that no bits of stone can give. It is, however, unusual for the general public to take much notice of these zoological finds, so the interest aroused when a strange fish was caught near East London, South Africa, some years ago, is rather difficult to understand. True, this particular fish had been reported missing, as it were, for 70 million years, but previous similar reappearances of other "extinct" animals had caused no sensation. For instance, in 1870 a lung fish was discovered in Australia (that home of antiques) of a genus that was supposed to have disappeared from this earth 170 million years ago, and the public could not have cared less. Yet when in December, 1938, a large fish of a bright-blue colour was caught by a trawler off Natal the event caused more sensation than an international football match, and photographs of the fish appeared in most of the world's newspapers.

The fish was one of the catch of a trawler brought in to East London, South Africa, where it was noticed by a Miss Latimer, the curator of the local museum, as a species quite unknown to her. She had it taken to the museum and stuffed, and she made a sketch of it which was sent to Professor Smith of Grahamstown who iden-

Coelacanth

tified it as a coelacanth of a type supposed to have been extinct for 70 million years. He named it Latimeria in honour of its discoverer.

The coelacanths first appear as fossils in the upper Devonian period about 300 million years ago. They died out (as it was thought) 230 million years later. They are large broad fish about five feet long, and their fins (with the exception of the first dorsal fin) are placed on the ends of what might be called arms. These arms possess a joint and muscles and undoubtedly represent the beginning of the development of the limbs of the land vertebrates. The coelacanths themselves are the descendants of the osteolepids I mentioned before, the group of fishes that are held to be the ancestors of mammals and man.

Over a period of 300 million years one would have expected considerable structural and anatomical changes to have taken place in the coelacanth, but the fish that was caught by the trawler in 1938 differed very little indeed from the coelacanth that swam the seas of the Devonian period, although shoots of the same stock during the same period changed so much that some of them now appear in the persons of ourselves.

Fossils of coelacanths up to the time of their supposed disappearance have been fairly common in various parts of the world, sometimes appearing together in great numbers. For instance, during the excavations for the Princeton University in America hundreds were unearthed from the Triassic rocks. That no fossils have been found for 70 million years shows that the absence of fossils must not be taken as proof of the extermination of a species. It does indicate, however, a big decline in numbers. Nor does the capture of a strange fish mean that it really is a strange fish: it only means that scientists have never come across it before. To native fishermen a fish is just a fish, something to be eaten, and to be eaten quickly in tropical regions where decay sets in in an hour or two. It is now known that coelacanths have been caught at more or less regular intervals off the east coast of Africa. Yet had Miss Latimer not chosen to take a walk around the docks that day, the coelacanth might still be on the list as a fish that died out long, long ago.

We will now return to the present story of the coelacanth. The specimen rescued at East London was "dirty and oily" and had been mauled about. Moreover, it stank. December is midsummer in South Africa and something had to be done. So the fish was stuffed. This was a good move and doubtless very necessary but it entailed the removal of the inside, and the internal organs of a newly-found living fossil are of even more interest to investigators than the framework. Professor Smith dearly wanted another specimen, so he circulated leaflets in English, Portuguese, and French, showing photographs of the fish and offering £140 for another catch—a sum that represented independence for several years to a native fisherman. Nothing seemed to come of it: natives are not great readers of leaflets nor changers of their ideas; moreover, the war intervened. Coelacanths may have been caught but if so the catchers evidently considered a fish in the hand was worth more than £140 in the bush. Anyway, no catches were reported. Then in 1952 just before Christmas, on more or less the same date that the first coelacanth had been discovered in East London, a telegram was received from the British owner of a schooner.[1] He said he had a coelacanth in the Comoro Islands.

This was good news for Professor Smith, but it would have been better news if the Comoro Islands had been nearer. They were two thousand miles away and again it was midsummer, and mid-

[1] This was Captain Eric Hunt who, in May 1956, was lost, together with his yacht, when sailing to the Seychelles in a reputed attempt to kidnap the exiled Archbishop of Makarios.

summer in the tropics at sea level is anything but keeping weather. How was he to travel two thousand miles in time to save the fish from complete decomposition? Another doubt gnawed at him. A seaman normally is not a scientist; *was* the fish really a coelacanth? There had been false claims before. One hundred and forty pounds engenders much wishful thinking.

But this did not stop Professor Smith from acting with great energy and enterprise. He persuaded the government to let him have a plane and when he got to the islands found that the fish *was* a coelacanth. It had been dead nine days but the British captain had embalmed it on the fourth day.

So a second specimen had been secured and in September, 1953, came a third. Houmadi Hassani, a fisherman, fishing very deep off Anjouan Island struck a large fish that put up a tremendous fight. It was half an hour before he could bring it to the surface, and his troubles were far from over then. In the end he coshed it on the head and landed it.

He felt sure it was what had now come to be called "The Fish," and hastened to the home of a Dr. Garrouste, an authority who had been provided with a preserving outfit against just such an occasion, by the Scientific Research Institute of Madagascar. So Dr. Garrouste was wakened at about 2 A.M. by an excited Hassani who said he had caught "The Fish" and that the doctor must go with him to see it. Not unnaturally, the doctor was not too keen to lose his night's sleep and take a long walk, particularly since several fishers had dragged him out before to view specimens they had caught that were not even remotely like coelacanths. He questioned Hassani. The fish, said Hassani, was brown with white spots and had eyes that glowed in the dark. Now the doctor knew that a coelacanth is blue all over and does not have luminous eyes, so he sighed and told Hassani to go back home and forget all about it. But Hassani refused to go unless the doctor went with him, and in the end the doctor went and when he saw the fish realised, in spite of its being brown with white spots and luminous-eyed, that it was a coelacanth.

It was sent to Professor Millét in Madagascar and by the time it got there it had turned blue and its eyes had lost their luminescence. Hassani got his £140 and a public ceremony as well.

Four months went by, and then at midnight on the twenty-ninth of January, 1954, the administrator of Great Comoro Island was wakened from his sleep by the arrival of the fourth coelacanth. This time everything was ready and the administrator and his assistant worked until morning injecting fluids into the great carcass.

They had not quite finished when another fisherman arrived with the fifth coelacanth, a specimen considerably larger than the other. Back to work went the weary operators, though the first fine frenzy had departed. Two days later the sixth coelacanth was brought to them and by that time they were definitely growing tired of coelacanths.

Preparations were made for dealing with a large influx of these fishes but no more were taken for eight months. (September to February would appear to be the "fishing season" for coelacanths.) Then coelacanths began to arrive once more. One of them was captured alive in November, 1954, and put into a whale-boat that had been filled with water. It was nighttime and the fish appeared to feel no ill effects from having been hauled up from a depth of nearly one thousand feet, but swam about in a normal manner. When the sun rose, however, it grew ill at ease and tried to hide in what shade there was. At the depth from which it came it is always dark, so to put it in a few feet of water in a tropical sun seems little short of madness if the intention was to keep it alive.

Professor Millét arrived at midday, but in any case, it was too late by then. The fish was still swimming, but seemed to be in agony. Its efforts grew feebler, and about four in the afternoon it turned belly-upwards, gasped for some time, and died.

There was excitement about this fish because it was a female. A female had long been wanted for it was thought that if it contained eggs these eggs might throw light on the previous history of coelacanths and on their descent. This female, however, carried no eggs.

In all, at the time of writing, eleven coelacanths have been caught.

Coelacanths live at depths of about five hundred to one thousand feet, a fact that has probably contributed to their survival. The heaviest brought in for examination so far weighed 130 pounds, but native fishermen state that this is far from being a large specimen and that fish of 225 pounds have been caught in the past.

The anatomical findings of the experts have not yet been given in full.

By the time this book is in print there will probably have been several more coelacanths taken, but no great haul can be expected of fishes living at such a depth. In any case, the coelacanth has long since ceased to be "news."

CHAPTER 8

THE CRUSTACEANS

The SEA mammals and fishes dealt with in the previous chapters all belonged to one phylum, the vertebrates. A phylum (as you doubtless know) is the first and broadest scientific grouping of life after animal and vegetable have been sorted out. A vertebrate, as its name implies, is a creature possessing a backbone, and so amongst the vertebrates we naturally find man and the other mammals, birds, reptiles, amphibians, and fishes, a very distinguished company, and most of them easily recognised. But for some reason (and the reason is complicated) creatures like sea-squirts, those stationary globules of jelly, and certain sea worms that burrow in the sand have been pushed into the vertebrate group although they are invertebrate. So whales and men have to belong to the same club as sea-squirts and certain selected worms.

We now leave the vertebrates and enter another phylum, Arthropoda. It is a more lowly phylum but containing an infinitely larger number of members. It also contains, on land, most of man's worst enemies, the insects and other small forms that attack man, his crops, and his domestic animals. But we are dealing with the sea, not the land.

Phyla are divided into classes and one of the classes of the arthropods is called Crustacea, nearly all of whose members (the wood-louse and one or two others are exceptions) are sea creatures.

There are probably more crustaceans in the sea than all the other animal life there put together. The species are mostly small and the class has been called "the insects of the sea." But many of them, in fact most of them, are much smaller than any insect, though on the other hand there are several that grow to a size larger than any insect can hope to attain, the giant crabs and the lobsters heading the list.

For some reason we are inclined to reverence size. A couple of heavyweight boxers attract more gate-money than a couple of featherweights, in spite of the fact that the featherweights usually put up a better show. Similarly, we are interested in those crustacean heavyweights, the crab and lobster, yet the semi-microscopic copepod is of far more importance than they. Were all crabs and lobsters to disappear it would mean nothing, but were copepods to go every animal in the sea would be affected, while on land, man, that great fish-eater, would have to draw in his belt several inches, for his fishing fleets would return with negligible catches and the great shoals of herring, cod and the rest would be reduced to a few brigades. In short, famine, so common on land would, for the first time in its history, come to the sea. Things would adjust themselves, but the whole sea population would be reduced. Copepods, of course, are not the only creatures, nor the only crustaceans, that feed on the diatoms and are the first process in the gigantic food factory of the sea, but they are the most numerous and most important. Indeed, they are the most important animals in the world.

A crustacean means simply an animal that has a crust. The crust is usually thin, though certain members, such as lobsters and crabs, by borrowing lime from the water have made themselves very hard and thick external suits. The number of species of crustaceans is legion, and most, as I have said, are very small even though they possess long scientific names. The habits of most of them are not by any means fully known. I shall select, therefore, for a short description only some of those crustaceans that are known to all of us and have popular names. All of them are edible, which adds to their popularity, but probably all crustaceans are edible; the trouble is they are too small to eat, unless one possesses the filtering apparatus of a whale or a basking shark or a herring or mackerel and can consume them as a sort of broth.

Pride of place will be given to the ubiquitous crab, the best known of all the creatures on the shore.

I once read the answers to a general knowledge paper set for children. One of the questions was: "Say what you know about (1) a beech tree; (2) a cuckoo; (3) a crab." Much ink had flowed over the cuckoo, but without exception all the class knew about the beech tree was that it was a common kind of tree with specially shaped leaves. This seemed to me a poor effort on the part of the class until I realised it was also practically all I knew myself. The crab ran a close second to the cuckoo and its description in most cases was good. Rather monotonous became the personal experience theme, "Once whilst fishing from the jetty I felt a tug and, thinking I had caught a fish drew in the line, only to find . . ." etc. Embittered, probably from a recent holiday experience, was the boy who wrote, "The crab is found in pools on the seashore. It is very vicious and inflicts dangerous wounds, biting to the bone with its claws." A slight misconception, however, prevailed generally. All of them seemed to think that there was only one species of crab, which, when small, frequented pools and shallow water and when grown-up went into deep water and became the crab that people eat.

There is, of course, almost no end to the species of crabs in the sea, some minute, some comparatively large, but all more or less of the same shape.

Crabs are engaging little creatures. Everyone (except the boy in the exam) likes them. Their comical sideway runs, their strenuous though usually inadequate efforts to escape capture, their pathetic threatening attitude when cornered, their vitality and zest for life and refusal to "give in," and above all, the fact that they cannot run so fast as the small human beings that so often pursue them —these things endear crabs to human beings though nothing will endear human beings to crabs.

As I say, the species of crabs are many. The majority live in the sea but not a few live on land and in fresh water. The fresh-water crabs we will dismiss; they do not concern us, but we cannot dismiss the land crabs, for practically all of them spend their early lives in the sea. Indeed, in land crabs there is taking place almost before our eyes another invasion of dry land by sea animals. It is a minor invasion and started much later than the famous fish invasion, but it is being run on the same lines. The shore crabs that children pursue over the sand represent the amphibians of old; they are converting their gill chambers into lungs, and some species that now reside farther from the sea have gone so far with this conversion that they will drown if kept too long in water.

Land crabs exist in many parts of the world, being particularly abundant in some of the Indian Ocean islands where they live in the hills. They are normally rarely seen, being nocturnal and far more cunning than rabbits in making holes and hiding themselves. Every year, however, they descend from the hills in their annual trek to the sea. They come down in battalions, their armour rattling as they move. They are large and look sinister. One observer described these massed scuttling armies as horrible and nightmarish. They surmount all obstacles like locusts in the hopper stage, and even overrun houses. They are bent on no aggression, however, and except for the nervous system do harm to no one. They have one overpowering urge, and one only, to get to the sea they have not seen for nearly a year. Clusters of eggs are attached to the bellies of the females, and in the sea these hatch out, some being washed off first and others hatching where they are. About two weeks later the crabs emerge from the sea in twos and threes and return to their hilly haunts or holes in the sand to resume their inconspicuous existence.

Meanwhile the young are passing through their various changing larval stages in the water, and when this has been accomplished and they look like tiny crabs and not something quite different, they, too, emerge and cluster in thousands on the rocks.

When the egg of a crab hatches a speck emerges that moults within an hour and turns into a creature that bears no resemblance whatever to a crab. It is about one-twentieth of an inch long, translucent, and carries two long spears, one on the middle of its back and the other projecting in front, like a beak. It has large eyes, set flat and not on the tips of stalks like those of its parents. It swims actively and feeds on small life on the sea surface. Other moults take place during which the baby crab presents an astonishing variety of completely different shapes. Finally, however, it loses the gift of swimming and sinks to the sea bed, still exceedingly minute but by now a replica of its parents in every way, stalked eyes, pincers, and the rest. Moult after moult takes place until it reaches adult size. This is the general run of things with sea crabs, but most

fresh-water crabs and some sea crabs emerge from the egg microscopic in size, but in other respects exactly like adults.

Amongst the sea crabs proper the sizes range from that of small
pebbles to that of *Macrocheira kæmphferi*, the Giant Japanese
Crab, which has the distinction of being not only the largest crab
but the largest crustacean. This is one of those long-legged crabs.
Its body is only about a foot and a half long, but when this crab
is put on an ordinary table the legs more than reach the floor, the
longest being about nine feet. The pincers, for the size of the creature, are insignificant. I met this monstrosity often in Harbin in
Manchuria, though only in a comatose condition, or served at table.
We called it the Vladivostok crab, and all the specimens came by
rail from Vladivostok. For some reason we never used the body, but
the legs, though they would have been called spindly in a crab of
lesser size, contained enough solid and excellent flesh to feed six
people. Once in Vladivostok a newly caught, live, half-grown specimen was put in the bedroom of a newcomer to greet him when
he returned from the bar. It was a callous joke and undoubtedly
took years from that young man's life. How many more years would
have been taken away if the crab had been fully grown I do not
know.

All sea crabs are edible—at least I know of none that are not—
and all, I think, are equally good. I have tried many. There was a
time, indeed, when I boiled and ate any crab of reasonable size I
could get hold of and I got the impression that many of the unusual
crabs tasted better than the conventionally edible varieties. But
with food, as with dress and almost everything else, mankind is conventional and eats what the others eat. Thus there was a time in
Britain when the ordinary Shore Crab (*Carcinus mænas*) was *the*
crab to eat, and few would look at *Cancer Pagurus* which is now
known as the Common Edible Crab and which appears exclusively
on the fishmonger's slab, and were a fishmonger to present shore
crabs for sale he would probably be regarded as homicidal. Yet in
certain places on the Continent it is still the shore crab that is regarded as the best of the crabs and that commands the highest price.

In America the blue crab is the "edible" crab but the Americans
are not so hide-bound as those over the water and many different
species of crabs are used. Of course to those who have not been

introduced to it, the idea of eating a crab at all is unthinkable. A fisherman in Ireland who had been using crabs for bait all his life refused to believe that the large ones he got in his traps could not only be used as human food but were of much more value than the fish they caught.

Most crabs cannot swim, but walk or run about on the sea bottom, eating molluscs or anything they can get, which in view of the tremendous competition both from their own kind and others cannot be very much. Small wonder they eat corpses and decaying stuff when they can get any. Crabs cannot afford to be particular; they live hard lives. But there are crabs (mostly belonging to the family Portunidæ) that can swim. The famous blue crab of America is one of these. Swimming crabs swim by means of paddles at the ends of their legs and a lot of them can swim and dodge almost as well as a fish.

The last creature from which one would expect intelligence is a crab. Yet if one judges by behaviour, and it is difficult to think of any other standard, certain crabs possess considerable intelligence and cunning. It has been said that men are the only animals to have learned to carry weapons, offensive or defensive. The monkey may hurl a coconut from the top of a tree but it never carries a stick. Man, however, was *not* the first creature to carry weapons; the crab had been doing this long before man reached even the ape stage. To what extent the crab knows what it is doing does not concern us: it *does* it. It will be said, of course, that only those crabs survived that carried these weapons and so the process became automatic and instinctive. But the many species that did not carry weapons also survived. The nearest analogy comes from a number of the same Arthropodic group, who hit upon an even cleverer device—the web-weaving spider. The actions of both are instinctive now; it is the origin of these schemes that give rise to thought.

To appreciate the subtlety of our first example, a crab named Dromia, we must first examine sponges. To us a sponge is one of the softest things in nature. In a bath we hardly feel its friction as it caresses us and douses us with warm water at the same time. It is also probably sweet-scented from some lavender or previous bath salts. In nature a sponge is nothing like that, it is covered with fine needles of lime or silica. It also has an offensive smell. Consequently it is given a wide berth by all forms of life except worms and other small creatures who live securely in its tubes. The crab, Dromia, takes advantage of the sponge's unpleasantness and converts it to its own use. It takes a living sponge, cuts and trims it to the size of its own back, places it on its back, and holding it down with its last two pairs of legs proceeds about its business both camouflaged and protected at the same time. It is thus rather like a woman holding a large hat on a windy day and has to keep two pairs of legs permanently employed clamping the sponge down, but the other legs suffice for its other affairs and the protection the sponge gives amply compensates for their absence from normal duties.

Other deliberate camouflagers are the spider crabs. In *Macbeth* an outpost watch of that erring but harassed king reported to his master that he thought he had seen the wood begin to move. Similarly, peering into the water you may possibly see a patch of seaweed and other trailing growth begin to move. This, in all probability, is a spider crab and the spider crab has done exactly what Malcolm's soldiers did; it has cut off vegetation and placed it over itself as a disguise. "Nonsense," you may say, "seaweed and other marine growths fasten themselves on to anything at the slightest opportunity. That crab simply happened to have got overgrown with the stuff without knowing it." You are wrong. The seaweed was deliberately planted by the crab. To prove it, catch one of these crabs, remove all growth from it and place it in an aquarium amongst growing seaweeds. Here you may observe it pluck pieces of weed and place them on its carapace where they are held by the crooked hairs until they "take root."

But that is not the end of the spider crab's repertoire. Take this same crab with its trailing garden attached to it and put it in an aquarium that is full of vegetation or other growth of another kind, small sponges maybe, anyway, things different. Here, in this different environment, the little forest on its back is more an advertisement than a camouflage, so the crab picks off every bit of weed it carries and implants instead bits of the growths amongst which it finds itself. Such cleverness is rather bewildering, and it takes place with one of the commonest of our crabs.

After this it may not come as much of a surprise to hear that some tropical crabs carry over their backs (held also by the last two pairs of legs) the shell, or rather half-shell, of a bi-valve mollusc such as an oyster or a scallop. Under this accumulating evidence man's claim to be the first (and only) animal to carry weapons grows thin. Size for size, an oyster or scallop shell is a much better shield than those flimsy affairs the Norman knights were wont to carry.

So far we have only seen clever crabs holding or growing things over their backs for concealment or protection, or both. We have not seen them deliberately *carrying* weapons in, as it were, their hands. We shall now see them doing this.

The beautiful sea anemone is a peculiar plant. It is peculiar first of all in not being a plant at all but an animal. It is peculiar in the second place in being avoided like the plague by practically every creature that swims the sea. Yet it is soft, thick-stalked, and fleshy and the fact that the sea slug eats it (probably the only animal that does) shows that it cannot be poisonous—at least to some organisms. It is also fairly defenceless; its tendrils sting but that surely ought not to worry the larger voracious hunters of the sea. The anemone, of course, is a voracious hunter itself, swallowing and consuming anything it can trap, but its prey is small and it cannot go out searching for it, being anchored for life wherever it may be.

I was wrong really in saying the anemone was fleshy; its stalk and massive head looks so, but the whole animal is chiefly water. It is said that if one takes one of those gigantic jellyfishes, such as the Portuguese man-of-war (a yard or so across), and dry it off gradually, the final residue will amount to nothing more than a stain such as might have been made by a bit of grease on a paper. The same is said of the sea anemone. I have never made the experiment so I do not know, but the anemone in spite of its thick robust-looking limbs is certainly mostly water, so it is difficult to understand why it should be dreaded. Yet according to many naturalists it is hated by sea creatures even more than the sponge.

The unpopularity of the sea anemone has not been overlooked by the crab, who, as usual, has turned it to its own advantage. So we find in warm seas certain crabs that possess a pair of claws too small for defence plucking off with each claw a small sea anemone which they hold and carry about wherever they go. Nothing could be better from the point of view of protection and possibly aggres-

sion also, but the crab benefits still further from the "flowers" it
carries. The anemone, like the crab, is a hunter and lays hold of
anything that comes within reach of its tentacles. Much of this food
the ungrateful crab will detach and convey to its mouth with its
first pair of walking legs. However, the anemones probably get all
they need and their newly-acquired mobility puts them in a position to pick up more than if they were stationary.

A certain very common hermit crab, *Lupagurus prideauxi*, living
in a shell, invariably has an anemone attached to the shell in such
a way that the mouth of the anemone is just underneath the opening of the shell, protecting the crab and also being in a position
to take in any food the crab may drop. Both, in fact, get bits of
food from each other. When the crab grows too large for the shell
and moves into another it detaches the anemone from the old shell
and puts it on the new. In most cases, however, the anemone saves
the crab the trouble of finding a new house at all: its growing roots
gradually surround the shell and extend beyond it. They also dis-
solve the shell itself so that the crab finds itself in a roomy house
provided entirely by its partner.

Other hermit species such as the deep water *Paguropsis typica*
have abandoned a shell altogether and live in a tent formed of
anemones, which they carry about. Hermit crabs also often manage
to get into sponges, together with their shells, thus dispensing with
the need of finding any further accommodation, for, as with the
anemone, the sponge gradually eats away the shell and provides the
crab with a safe and comfortable, if malodorous, cavity for life.

There would almost appear to be some sort of understanding between crabs and the generally detested anemones. We are accustomed nowadays to view most associations of this kind in terms of
the word *parasitism*. In that case, is the crab a parasite of the
anemone, or the anemone of the crab? The crab uses the anemone
for protection, for a house, and takes part of its food. But the
anemone feeds also on part of the food of the crab and derives
benefit from being carried.

Actually, of course, neither the crab nor the anemone is a parasite. That word is far too loosely used. A true parasite is a creature
that lives on the tissues or juices of its host. Aphids are parasites

on plants, red mites on poultry, fleas and lice on men, and so on, but no further.

So we must find other words for other kinds of associations. There are two in common use, *symbiosis* and *commensalism*. Here again there is no strict interpretation of these words: at least they are not always employed according to their strict interpretation. The word *symbiosis* is used for animals that join together and are mutually dependant on each other. The Oxford Dictionary says "Permanent union between organisms each of which depends for its existence on the other" (such as fungus and alga composing lichen). But the word is often used for an association by which each party benefits though not necessarily to the extent of being unable to live without the other. *Commensalism* means "dining at the same table" or living together and partaking of the same food. So where to put the crab-anemone partnership is difficult to say. They are not strictly commensals, for in the main they have their own separate sources of food, and they are not symbiotic, though they benefit to a certain extent from each other.

Which all goes to show how inadequate words can be. Take the tapeworm. Who would hesitate to call that a parasite? Yet strictly speaking it is not a parasite but a commensal. It does not feed on the tissues of its host; it feeds on the same food that he does.

And here is another example. A tiny crab, known as the oyster crab, lives inside oysters, mussels and the like. It is always referred to as a parasite. It is not, it is a commensal, doing no harm to the oyster but merely, together with the oyster, feeding on the microscopic food that is washed through the oyster's opened shells. According to the ancients the association was even more involved, for they believed that the crab warned the mollusc of the approach of enemies so that the mollusc could snap tight in time, thus protecting both itself and the crab. So if the ancients were right (and they were not always wrong), this association verges on the symbiotic.

We have admired the artfulness of the hermit crab in using anemones, but the normal scheme of life of the ordinary hermit crab is clever in itself. Or perhaps I should say "appears to be" clever, in the fond hope of warding off indignation from those who take the word too literally. All creatures must have some device to cope with competition and conditions. Man, for instance, was in a precarious position until he learned to live in a cave and carry a stick with a stone tied on the end. He gradually improved his stick until it blossomed forth in the shape of an atomic bomb, and if he finds he has carried his improvements a little too far he will not be the first form of life to suffer owing to over-elaboration of protective or aggressive devices. Every living creature in the sea and on land has its own methods, but some methods are cleverer (I insist on using that word) than others. We see hermit crabs every day by the sea, and their habits have grown stale, but were they some rare species written about for the first time by an explorer, we should either marvel at this new instance of cleverness in an animal, or disbelieve the explorer. Many animals find homes to hide or live in, but few acquire protective homes that they can carry about with them.

The rear portion of the hermit crab is soft and so fashioned that it fits into any curved shell of the right dimensions. Its front part is armoured, so that the crab is protected both behind and in front and at a moment's notice can retire into a fortified castle, and then, the next morning, pop out and run about and hunt as actively as if it had no castle trailing along behind.

Alas, however, the hermit crab is subject to the curse of all crustaceans, it must moult. Its shell becomes too small and it must find another. It is very loath to leave the old shell, not for sentimental reasons but because when it does so its soft hindquarters will be at the mercy of hundreds of voracious feeders, including its own kind. Indeed, to the crab it must seem that all the hungry world is waiting for a hermit to change its quarters. Fortunately it is able to go house-hunting whilst still occupying the house it proposes to leave, so that when it does find a house of the right size the change-over is quick, though often not quick enough. It has been roughly estimated that about 50 per cent of hermits lose their lives when changing quarters. First of all, other hermits are probably seeking new abodes, and there may be a free fight in front of a suitable shell; secondly, the new shell must be investigated lest some other occupant is in residence, such as a worm or worse. During this brief investigation another hermit may have run off with the owner's original shell and the owner, expelled by an indignant tenant, may find himself homeless, a condition unpleasant to most creatures but fatal to a hermit crab. Even without complications the scramble from one shell into another is a process as dangerous to the crab as dashing across a busy road, without looking, is to us.

Many hermit crabs have taken to land. Here mollusc shells are not so plentiful as they are in the sea and some hermits have taken to wearing broken coconut shells instead. Indeed, I have heard of one that was seen in occupation of an empty soup tin. Others on land have abandoned shells altogether, and so in the course of evolution have finished up where they started, shell-less. This freedom has been attained by a hardening of their posterior regions.

From the point of view of the human gourmet the lobster is the aristocrat of the crustaceans. He is blue-blooded too; the blood is almost colourless but it *is* faintly blue. There is something plebeian about the crab and he can only hope to appear at the best tables in disguise, whereas the lobster, as nature made him, though cut in half, is the crowning glory of the salad dish. Unfortunately for themselves both crab and lobster have to be boiled alive before they become of any interest to the majority of human beings, but this worries no one. Any housewife who would go into agonies of remorse if she trod on the cat's tail will perform this operation without a qualm.

The usual way is to put these creatures into cold water and bring it to the boil, which seems to be prolonging the agony, but it is said it is "kinder" to start with cold water than to plunge them into boiling water, the rapidly changing temperature killing them before the water gets really hot. That is as may be, but whichever process is the "kinder" the big caterers now always kill their lobsters and crabs before boiling them, which is "kinder" still. This is not done from humanitarian motives but because these crustaceans, particularly crabs, are apt to kick several limbs off, including the pincers, in·their dying spasms in the pan. For the ordinary housewife or hotel chef, however, the difficulty is to know how to kill them. Amongst the electric gadgets continually coming into use in the kitchen there is room for one for electrocuting crabs and lobsters.

The outer shell of the lobster is a murky blue colour which, on boiling, changes to a brilliant red. Quite how this comes about I do not know but I dare say any chemist will be able to tell you.

This transformation is a lucky one for the fishmonger. For some reason red excites gastric desires, not only in flesh but in other foods: a crimson-cheeked apple will always sell better than a greenish one of infinitely better flavour (sprays are now in use, not for controlling apple pests, but for reddening the skins); and all efforts to popularise the sweeter yellow tomatoes have failed; the reds win every time and always will, on account of their colour. Did a lobster turn green on boiling or even retain its original leaden colour, we should have to overcome a natural bias before we bought one.

Still dwelling on lobsters from the food angle, the best known and most widely used are the American Lobster (*Homarus americanus*) and the European Lobster (*H. gammarus*). The anatomical difference between these two species is so slight as to need an expert to detect it, but the American lobster grows to a larger size— twenty-five pounds, possibly, for an American against ten pounds for a European, though you are not likely now to meet giants like these on your normal shopping rounds.

In North America the lobster is *really* important. The canning industry alone supplies the world with lobsters and is big business. Over-fishing has resulted, though this is apparent not so much in lack of lobsters as in a decrease in size. The large lobsters that used to appear at table are no longer to be found. The preservation of lobsters generally is assisted by minimum-size limits in the catch (the increase in size limit from nine inches to ten and one half inches in Maine made flourishing what had been a dying industry), but this is not the full answer. It is the big females that are the real breeders and they spawn an infinitely greater number of eggs than the smaller ones; a lobster in its first season spawns about nine thousand eggs, while a large female spawns one hundred thousand. Attempts have been made to institute a closed season, but the female lobster makes this difficult by carrying her eggs about with her for nearly a year. A few years cessation of lobster fishing altogether would work miracles, but this is unlikely to come about.

There are of course other edible lobsters in all parts of the world, including the sea and fresh-water crawfishes that carry no pincers. Edible is the wrong word, for the lobster that cannot be eaten does not exist. Some, however, are better in this respect than others and the small Norway lobster, sometimes called the Dublin prawn, takes a lot of beating. It has slender inadequate front claws longer than its body, which is about five inches long, and it is very prettily coloured even without boiling—orange, with red and blue markings. Unlike its larger commercial relatives it lives in deep water (fifty fathoms or so), and on a mud bottom (most lobsters prefer a rocky bottom), so it has to be caught by trawling and not in pots. It is never, so far as I know, trawled for deliberately but comes up with the fish catch and since it has to be boiled almost immediately if it is to go to market, we rarely see it on the slabs—which is a pity.

Everyone has seen a lobster hundreds of times, so it may seem superfluous to describe one. Moreover, none of the readers of this book are studying the crustacea for an examination, at least, I hope not, for my casual survey would not earn them many marks. All the same, I should feel a slight prick of conscience if I passed by the crustacea without a description of one of them, and a lobster is as good a general type of crustacean as any. My description will be cursory; the more intricate details will be left for the dissecting class.

The appendages in front of the lobster that rather strike the eye, the long feelers and other protuberances, are by name the antennæ, the antennule, and the maxillipeds, and an account of their functions, or supposed functions, would commit us to too much paper. Over the front part of the lobster, the back, or cephalothorax, lies a thick armoured cuirass called the carapace, in shape and fitting not unlike one of those jackets put over toy dogs to protect them from the cold. The owners of these dogs might take a tip from the lobster and make the jacket proof against enemies as well. Jutting out in front of the carapace, and a part of it, is a long narrow peak called the rostrum. This passes over the eyes and probably serves to protect them. Incidentally, it is only a small difference in the protuberances on this rostrum that distinguishes the American edible lobster from the European.

Underneath the carapace, at the lower part, are five pairs of legs. The front, or first, pair of legs are not used for walking and it seems rather odd that they should be classed as legs at all, for they are what we call the pincers, those massive claws that give us meat when we crack them. Their scientific name is the chelapeds.

Closely following the chelapeds come four pairs of real walking legs, though even these are not used solely for walking; the first two pairs have small pincers at the end and can perform feats not unlike that of a clumsy hand, such as picking up selected bits of food and conveying them to the mouth.

The claws, or pincers, of the lobster do not grow in proportion to the rest of the body. As the lobster increases in size the claws increase also, but at nearly twice the rate. This is only one of many reasons why, gastronomically, large lobsters are preferable to small. In the miserable specimens we get today the claws are so small that many do not think it worth while bothering to crack them to get the meat. But consider a large lobster whose claws have had a chance to get really going. The largest lobster ever known was an American one that weighed thirty-five pounds and the claws alone of this gigantic creature weighed twenty-three pounds. Specimens like this must be well fitted to cope with life as they find it in their particular sphere. In their cumbrous armour they move slowly, but food in the shape of molluscs is everywhere and those pincers of theirs can crack open any mussel or clam or scallop they come across. Their armour, too, at this stage must be proof against almost any natural enemy. But man has seen to it that they no longer exist, except perhaps in aquariums.

Going back to the normal-sized lobster we are examining, you will of course have noted that the two claws are different. One is smaller and has the inner edges serrated like a saw, the other, the large one, has a few blunt knobs on the edges. The former pair are used for tearing prey or meat into small pieces, the other for breaking the shells of molluscs. Incidentally, both can be equally punishing to the human finger. I have not said on which side these different claws are situated, because they may be on either side; one specimen will have the big claw on the left, another on the right. This does not matter, but it is surprising that Nature, usually so meticulous about her anatomical arrangements, should be so casual in respect of the lobster.

Food is usually conveyed to the mouth by the second pair of legs, and the lobster is more of a believer in thorough mastication than ever Gladstone was. It has six pairs of working jaw appendages which reduce the food almost to pulp before it passes down the narrow gullet into the stomach, and in the stomach itself there are teeth-like protuberances that grind up the food still further. The lobster can never suffer from the indigestion it so often gives others who do not follow its own precepts of thorough chewing.

Lobster

If you seize a lobster in the sea by a leg or a claw and hold on, that leg or claw will break off and the lobster will scuttle to safety, leaving the severed member in your possession. Now in higher animals the wrenching off of a limb will cause death through bleeding, and, but for a special arrangement, death from this cause would happen to the lobster also. But when a limb is lost, the fracture is always made at a certain joint near the base, where there is a very narrow passage through which pass the congested nerves and blood-vessels. So small is the aperture that, after fracture, a clot of blood can form and block it completely. Should, however, the breakage occur at any other point, or should a limb be crushed and mangled, then the lobster *will* bleed to death unless it can manage to throw off the remainder of the limb at the proper place and get a blood clot.

The lobster does not remain limbless long. A scar, or bud, forms at the point of cleavage which grows bigger and which at the next moult takes on the form of a small replica of the lost limb. This becomes larger at each moult until finally it is indistinguishable from the original. Incidentally, Nature occasionally slips up over this replacement routine. If, for instance, the large crushing claw is lost, Nature may replace it by one from the wrong drawer and supply a saw claw instead, which would be tantamount, in our case, to receiving a pair of left hands.

The eyes are at the base of the rostrum, or armoured peak, and would see little were they not perched on movable rods like periscopes and so are able to see either in front, behind or sideways. Each eye has thirteen thousand lenses, and a nerve rod is connected to each lens. Whether an eye with this number of lenses is thirteen thousand times better than an eye with a single lens like our own is more than doubtful. Much has been written about the sort of image conveyed by these compound eyes, but unless we can become a lobster or a dragon-fly we shall never really know.

It is rather mortifying for us men that although we have managed to control nature to a certain extent, we have no physiological control over ourselves; we have to put up with what nature gives us. And nature does not seem to realise our importance. I have already commented on the injustice of giving replaceable teeth to fishes and reptiles but not to us. Now we find that she has given replaceable limbs to mere lobsters and other crustaceans but again not to us. Yet a lobster has ten legs against our two. And there is worse to come. Nowadays artificial teeth and limbs can to some extent replace missing members, but there is no substitute for an eye. Cut out a man's eyes and he will never see again, but cut off a lobster's eyes and nature will grow new ones for him, complete with the thirteen thousand lenses and thirteen thousand nerve rods, and as good in every respect as the originals. There is only one proviso: you may cut the eye off from anywhere down the stalk except at the extreme base. Here the nerve ganglia is situated and if that is destroyed no new eye will be formed, and the stalk will merely develop into a kind of short and useless antenna.

The carapace is a thick and hard piece of armour-plating, but it is different from the steel cuirass worn by a guardsman or the shell of a mollusc in being sensitive. Fine hairs that are connected with nerves underneath grow from it so that the lobster feels a slight touch on its carapace as readily as a man feels a touch on his bare skin.

I cannot say that I enjoy giving these anatomical descriptions, but plodding wearily backwards along the lobster we now come to the tail. This is more properly called the abdomen even though it is composed chiefly of solid flesh. A small intestinal tube, however, does run through the centre. It is protected by five plates (excluding a very small one next to the carapace) that slide in and out and allow movement. The allowance is generous, for when called upon the lobster can lash its tail most vigorously. The sliding plates, however, only move on one plane, therefore the lobster can move its tail up and down but not sideways. Normally the lobster does not lash its tail but walks sedately along the sea bed when not hiding in some cranny, but if an enemy is encountered the lobster shoots backwards at speed by its tail movements. This may sound an unusual form of progression, but it has its advantages; it gives a flying start, without having to turn round first; it keeps the enemy in full view during the pursuit; and it is the fighting forepart and not the defenceless rear that the pursuer must deal with if it manages to close in.

Under the abdomen, or tail, are five hairy paddles called swimmerets. The lobster's ancestors undoubtedly used these for swimming but nowadays they are next to useless for that purpose. They are still important, however, as egg oxygenators. When the female is in berry—and she usually is—she carries her bunches of eggs under her abdomen, the bulk of them attached to the swimmerets. Now, as we know, she often carries these eggs for nearly a year, but if they were just carried in a bunch like so many peas in a sack they would die from lack of oxygen. The female therefore sends oxygenated sea water (and all sea water is oxygenated) flowing through them by moving her swimmerets to and fro perpetually.

The tail fan comes at the end of the abdomen and with it we conclude our superficial examination of the outside of a lobster. The inside is more difficult, and again we will leave all the real work to the dissecting class, for the organs of crustaceans, though to a large extent fulfilling the same functions as our own organs, are not like them in appearance and are not co-ordinated in the same way. Still, we, in the first form, will probe hastily into the inside of a lobster, merely mentioning some of the various organs and showing where they are.

Inside the shelly front jacket we call the carapace lie most of the organs of consequence. First, as might be expected, comes the brain, situated, as with so many animals, in the head just behind the eyes. Immediately after the brain, rather surprisingly, comes the stomach. It is a capacious stomach, containing not only strong digestive juices but grinding teeth. Such a stomach would be worth a fortune to any dyspeptic human being. In the congested conditions inside the carapace, it seems to occupy more than its fair share of space. Situated also in front and at the sides of the stomach are glands that correspond to our kidneys. The only unusual thing about them is that they are bright green in colour. Indeed, the lobster is a colourful crustacean, inside and out. Behind the stomach is the liver, a large, yellow-green mass of soft tissue. And behind the liver (in the male) is the testis, which is connected by the genital duct to an opening between the last pair of legs. In the female in the same position, though occupying considerably more room (the female is broader across than the male), are the ovaries, well known to every lobster-eater as the "coral," for on boiling they take on a cerise colour.

It may have been wondered when we were going to come to the heart; we come to it now, the rearmost organ in the carapace and lying over the testis. Were our own anatomical arrangements similar to those of a lobster, actors and opera singers would have to place a hand over the pit of the stomach to indicate devotion.

Mention of the heart brings us to the heart's work-mate, the lungs. The lungs of a lobster are gills, but they perform the same function—the collection of oxygen.

They are not, however, similar in appearance to the gills of a fish. I will not go into a minute description, for everyone has seen them, if not in a lobster at least in a crab. By everyone I do not mean diners at restaurants, for if such see them in the crab or lobster served they have every reason to complain to the management. I mean the ordinary man who occasionally takes home a boiled crab from the fishmonger. As he breaks it open his wife is almost sure to tell him not to forget to take out the "poisonous parts" and, being as averse to death as she is, he will not forget, but will remove a number of long, stringy, putty-coloured strands that curl round at the base from under the armour at the sides. These are the gills, and the fact that they are not at all poisonous does not matter; they are inedible and should be removed.

The heart, possessing valves for the control of the blood-stream, pumps the lobster's blue blood through every part of its body in many streams that finally join together at the dial's centre, the gills. The gills—we must be brief about what happens now for the subject is complicated and we do not wish to get seriously involved— the gills are full of fine channels separated only by the thinnest of membranes from the oxygenated sea water that passes through them. As it passes through these thin membranes, the blood takes up oxygen from the water and gets rid of its carbon dioxide. To operate, gills must have water continually flowing through them, just as our lungs must have air continually drawn in. The lobster has its mechanism for pumping water through the gills. Back to the heart goes the re-oxygenated blood, to be despatched again on its journey through the body.

The apertures of the generative organs are between the last two pairs of legs, and the female has a cavity there for the reception of the male sperm, which she retains until the eggs are ready for despatch from the oviducts.

I need hardly say that when the egg hatches, the larva looks different from its parents. It swims freely and possesses a row of spines down the middle of its back. But the difference is not so great as with the crab; indeed, with a little imagination, the larva can be passed as not un-lobsterish in looks; and soon it sinks to the bottom and moults and creeps about like the adults.

The food of the lobster, like that of the crab, consists of molluscs and anything else it can get, living or dead, preferably dead. Lobster-pots are usually baited with bits of fish, and those who know tell us that the more decayed the fish the better bait it is. But do not let this deter you from eating lobsters if you can afford them; their capacious stomachs can deal adequately with any food, however rotten.

In spite of their armour, normal-sized lobsters have many enemies apart from men, including fishes and other sea creatures, but *the* dreaded enemy is the octopus. The octopus, generally speaking, is not so black as he is painted but he is death to lobsters. The lobster, when wedged in its cranny, is secure from most preyers, but the octopus, itself a dweller in crannies, drags it out and holds it in its sucker-studded arms, inserts its beak in some chink in the middle of the under part, injects a paralysing fluid, and then eats the lobster by sucking out its contents. Normally, octopuses are resident mostly in warm waters but occasionally (one cannot be quite sure of the reasons) they invade colder seas, not in ones and twos but in large numbers. Such an invasion occurred in the English Channel several years ago and well nigh ruined the lobster fishery. It took many years for the lobster population to recover and there are those who affirm it never did.

I have mentioned the frequent shell-casting of crustaceans as casually as if not much more were involved in the operation than the removal of a pair of riding-boots or silk stockings is with us. Yet every time a mature crustacean moults (and it does so many times a year), it performs the seemingly impossible. We will take the lobster as an example, not only because its make-up is familiar but also because its method of moulting has been frequently observed.

There was once a man called Houdini who used to intrigue audiences by freeing himself from handcuffs and any other bonds. Any lobster, however, would have been in a position to smile patronisingly at his performance, for the lobster has to disengage itself entirely from its outside covering, including that of the long whiskers in front, the eyes, the legs, the claws, and he must (unless he slips up) so disengage himself that the outside skeleton remains just as it was before its tenant departed. How can the lobster do this? Consider the front claws alone, with the huge pincers and the narrow joints all enclosed in shell as hard as iron plate. We find it difficult enough to get the flesh out of these pincers even when we have cracked them and are using special probes, but the lobster has to get the inner claws out right to the base, and he cannot crack the shell. The same goes for all the other appendages. Confronted with a problem like this Houdini would probably have remembered an urgent engagement elsewhere.

However, the lobster is now on the stage (the stage is an aquarium) prepared to do his act (in as secluded a place as he can find). It is unfair, perhaps, but I will give away part of his trick before he starts. First, the blood leaves the limbs and all the appendages, and collects inside the body. Next, a general loosening of the inner portions from the outside coverings takes place. Then the shell begins to split down the thinner underside—the operation has begun.

It would all be in vain, however, if the flesh of a living lobster were like that of a boiled one. In a living lobster the flesh is not "set," so massive portions *can* be squeezed through minute passages, reforming afterwards to their original shapes. To get an idea imagine an inflated rubber balloon being squeezed through a small opening—not easy, but possible.

So eventually, if things go well, an entire lobster withdraws itself, leaving behind a shell that looks like another lobster.

In a museum in Germany before the war the cast shells of a single lobster that lived in a tank had been kept. There they lay, fifteen apparently normal lobsters ranged side by side, each one slightly larger than the next, the whole looking like so many different lobsters caught at sea.

Two major perils confront the lobster when it leaves its shell. During the actual moulting the loss of limbs is common and death may ensue, for not only the outer skeleton but certain inner organs such as the lining of the stomach have to be changed. Secondly, in a sea full of questing feeders, the lobster for some days is left, in the words of Shakespeare, naked to its enemies.

It is after moulting, when the female's body is soft, that mating takes place. The male conveys a special receptacle to the female consisting of a number of capsules, each charged with sperm. Two months later the eggs are laid (the female carefully cleaning her hairy swimmerets beforehand). According to observers the female then curls her abdomen into a circle and emits the eggs, which she coats as they emerge with sticky secretion. The male capsules, kept in readiness for two months, are now discharged at the eggs, burst, and release the sperm. The eggs, as we have already seen, stick in bunches to the hairs of the swimmerets and remain there for nearly a year.

Amongst the baggage of many small children going to the seaside will be found a shrimping net. Nothing will be thought wrong or peculiar in this by anyone, but should a grown man take one on *his* holiday, porters will smile and may even go so far as to ask him if he has forgotten his bucket and spade. No comments will be made, however, if he carries a fishing-rod. Yet there is no comparison between the two sports. The grown man will probably spend his time sitting down fishing from the pier or from a boat and if he hooks anything it is unlikely that its weight will cause any strain. Shrimping, however, is really a pastime only for strong men. If you have not already done so, try it yourself. Use a net of the proper size and push it, up to your waist in water, through sand and mud for two or three hours. After this, as you nurse your aching muscles back to normal, you will wish you had played water polo instead. The mental exercise is different also; the shrimper has no time to sink into the glassy-eyed coma that often overtakes the pier fisher. Even if he gets no shrimps, every few minutes his dredge will bring up a varied assortment of active life, and if he does not find this interesting then he ought to have stayed in his office.

I am not posing as an expert shrimper. When I go shrimping I like to go with my son who is strong and muscular and does most of the pushing, and it is a strange thing that when I do the pushing an odd shrimp is all I bring up, whereas he generally catches dozens. He also gets an occasional prawn but the ordinary shrimper is not likely to get many of these; they should be sought under rocky ledges where a shrimping net will not go. We were put to shame recently. It was one of those off days. Pushing our net along we saw a middle-aged woman with her frock tucked up bending down and beating the water with a stick at the base of a small rock. In her other hand she held a small net. We watched her for some time, but every time she brought up her net there was nothing in it. My son asked her what she was doing. She was prawning, of course. We commiserated with her and said that we, too, had caught next to nothing. Whereupon she opened a shopping-bag slung round her waist and containing quite a hundred large prawns.

Few people have really seen a shrimp in its natural environment. All one sees are greyish streaks and puffs of sand as the creature flies and hides. Even when lying in full view on the sand its semi-transparent body and faint brown markings make it practically invisible. It is one of Nature's best efforts in camouflage. Swimming in a pail of clear water it looks quite dark and well defined. It is conspicuous also when one lifts it up in the shrimp net, but conspicuous in a different way, for then it shines like a piece of glass.

There is a big industry in shrimps. In the U.S.A. the catch is about 70 million pounds a year, of which about 15 million pounds are canned. And many other countries are not far behind. Evidently, therefore, a lot of people must be fond of shrimps. So am I, but only when someone else does the peeling. It takes the average person a lot of time to get a shrimp out of its shell, and for what reward?—a few mangled pieces, at which one looks dubiously and wonders if they are worth eating. I knew a woman who with two or at the most three dexterous twists could get a shrimp out of its shell intact and she went so far as to show me how to do it, but I never learnt.

Shrimps and prawns, of course, are not confined to the sea, they abound in many rivers and lakes. In some inland rivers in China there is a very small species, superior in flavour to any shrimp I know. I used to have it at Chinese feasts that I sometimes attended when travelling in the interior. The Chinese, of course, possess inexhaustible patience and small delicate hands, but I do not envy the kitchen staff their job of peeling those almost microscopic shrimps which we devoured in large numbers. The rivers and canals where these shrimps abounded were in thickly populated areas. They were also the final home of dead dogs and other refuse even worse. So when eating the shrimps it was advisable to put all thoughts of what they fed on completely from one's mind. I repeat, however, they were delicious.

No one can have failed to notice how like lobsters prawns and shrimps are. They lack the exaggerated claws and have other small differences, but they are on the same model. Lobsters are probably descended from prawns and in the change-over adopted heavy armour, sacrificing speed for doubtful security. Enclosed in a covering not much more substantial than cellophane wrapping, shrimps and prawns can fairly shoot along in the water and twist and turn with the best. And they certainly succeed in life, but so, on the whole, do lobsters. We shall have more to say about armour later.

There are, of course, anatomical differences but, taking them by and large, there is no strict dividing line between shrimps and prawns. The small ones are called shrimps and the larger ones prawns, and going up the scale we get prawns that are really lobsters like the Dublin prawn. The Newchwang prawn of China, about ten inches long, might be called a lobster but its thin brittle covering is definitely prawnish.

The common prawn is about three inches long and frequents shallow water and the bases of rocks, and there was a time when this was the only prawn we ate. There are prawns, however, that live only in deep water. At one time the average man never saw these. Scientific expeditions brought some of them up and examined them and wrote about them, and that was the end. It is, however, due to scientific investigation that the number of prawns on the market has been almost doubled. Some time ago scientists made an intensive study of the deep-sea life of the Norwegian fjords. Their investigations were somewhat handicapped by the numbers of large prawns that were brought up in almost every haul. They did not want these prawns, they wanted to examine other forms of life, though if they had any sense at all they kept the prawns and boiled them for supper. The investigators departed, and it was not until some time later that the idea occurred to some commercially-minded individual to fish for these prawns and put them on the market. Now, deep-sea prawns from the fjords of Norway are the most popular of all prawns and are exported to many countries. Four inches long (six inches occasionally) they have both substance and flavour.

It is an interesting fact that many prawns emit light, for what reason no one knows. So do certain shrimps. The minute shrimp-like Nyctiphanes for instance (whose relative provides the bulk of the food for the whalebone whales in the Antarctic), when present in numbers can turn sea water into liquid fire. Night bathers in some parts of the world have been alarmed on leaving the water to find themselves apparently dripping flame.

Crustaceans, however, are not by any means responsible for all the luminescence in the sea. Other organisms, particularly flagellates, usually create the light in the water when disturbed by, for example, the screw of a steamer, the splashing of an oar, or even the waves or the falling of surf on the beach. This, incidentally, is often called phosphorescence, though since phosphorus has nothing to do with it, the word is wrong. On the other hand, we are ignorant as yet as to what *does* cause the light and why at certain infrequent times the organisms in the water, on disturbance, emit light of a particularly startling brilliance.

Last on my short list of familiar crustaceans comes the best known of all. The bulk of seasiders come away from the sea without ever having seen a lobster, prawn or shrimp. These crustaceans are well known only because their corpses are seen in every fish-shop. It is just the reverse with the barnacle: it is well known because it is on view every day by the sea, but not one will be found in a fish-shop.

Besides being the commonest of all shore creatures, it is also regarded as the least interesting: no more interesting, think most people, than the stucco-cement on a house. In giving this impression, and in other respects, the barnacle is deceitful.

The ebbing tide, like the rising curtain in a theatre, but going the opposite way, discloses other actors in the seashore tableau besides barnacles: massed mussels clustering on rocks and pier-supports, limpets and periwinkles. These, together with their kind, oysters, clams, scallops, whelks, cockles and the rest, are molluscs, and the most mollusc-looking of the lot are the barnacles.

Nearly all the crustaceans, large and small, have a shrimp- or crab-like consistency in their appearance, and move about, sometimes at speed. The barnacle is remarkable in many ways but the most remarkable thing about it is that it is not a mollusc. This news will probably not stir the reader, but it stirred scientists when, in 1830, J. Vaughan Thompson, after the very difficult task of tracking down the free-swimming larvæ of the barnacle, was able to show that the barnacle was a crustacean and not a mollusc—however much it tried to imitate the latter when it became adult. For in its early stages the barnacle, even to the not-too-expert eye, is obviously one of the crustaceans.

We will skip the various changing stages of the young barnacle and see it when it settles down in its last metamorphosis, and if

ever a creature settles down in life when it grows up, it is the barnacle. It falls from the sea surface where it was swimming and
feeding, and lands on the back of its head on a rock. In that position it immediately exudes cement to fasten itself down, grows a
thick shell, and stays put until old age or some parasite or whelk
puts an end to its inactivities. And so determined is the barnacle
to remain where it is that it goes on adding cement for the rest of
its life.

Its shell we know; rounded at the base with a kind of pyramid
on top which should be treated with respect, for more than one
holiday-maker has had the seat of his trousers torn by getting up
hastily from a barnacle-encrusted rock.

Inside this shell lies the barnacle on its head with its legs point-
ing upwards, looking from the outside merely like part of the rock
itself. This, however, is when the tide is out and the rock is dry
and possibly baking in the sun. When the tide returns and covers
the rock, if you could submerge yourself and watch one of the barnacles, you would see valves on the top of its shell open and a
feathery leg emerge (a "ghostly hand" is the description of one
writer). Soon the leg will begin to wave about, making a scooping
motion. This is the barnacle's way of feeding: with its feathery fan
it drives a current of water through its mouth (or what approxi-
mates to its mouth) where any passing microscopic food is intercepted and taken in. Soon the feet are withdrawn into the interior,
the valves close, and the barnacle is as it was before the tide came.
But only for a minute. Again the valves open, the limbs emerge,
and the scooping process is continued, and so it goes on.

This net-casting for food is completely blind. No calculating eye
lies behind the opened valves. The barnacle starts life with one of
the largest eyes possessed by any living creature, and finishes up
with no eyes at all.

But now the tide is again receding and disclosing the top of our
barnacle's rock. Steps must be taken. The barnacle can no more
live in air, without, so to speak, artificial protection, than a man
can live under water. It has to prepare for at least several hours
of dryness, and if it is situated near the limit of the spring tides
it may be several weeks. It has to undergo a prolonged siege under
dry and perhaps hot conditions where all the processes of a siege

except direct attack are brought into play—hunger, thirst, and, in the case of the barnacle, dessication and lack of oxygen (for the barnacle cannot take in oxygen once the sea has left it).

If you are amongst barnacle-covered rocks when the tide is leaving them and if you have moderately keen hearing you will hear a faint rustling like the sound of pellets gently shaken in a box. This is the slamming-to of innumerable little barnacle doors. As the receding waters lap against the barnacle it takes in a bubble of air, all the water it can hold, and then, like the closing of hatches in a submarine, it clamps down its four valves. These valves fit perfectly and are air- and water-tight. They have to be, of course, for any barnacle or species of barnacle that possessed imperfectly fitting valves would perish.

On the moisture and air taken in the barnacle must now subsist for varying periods and during varying conditions. As for food, it is to be hoped (from the barnacle's point of view, not ours), that it has scooped in enough food to last it over its period of stress.

Barnacles (unlike most crustaceans) are hermaphrodites, containing both sexes in one individual. Yet if alone they cannot breed; junction must be made with a nearby fellow. So at breeding time a tube emerges from one barnacle and delivers sperm to another. If no fellow is near enough for the tube to reach, the barnacle cannot breed but remains all its life, as Berrill puts it, a bachelor and an old maid at the same time.

There are about two hundred species of barnacles. Those one sees on rocks are generally known as acorn barnacles. Another type is the stalked barnacle. This attaches itself to submerged objects in the sea by a thick stalk and holds on with the same bulldog tenacity as its relative.

An authority once stated that although crustaceans were of no use to man except as an article of food in a few cases, on the other hand they did him no harm if one excepted a few land crabs that raided his crops. I will not quarrel with his first statement except to suggest that to provide food for man on the scale of the crabs, shrimps and lobsters (ask the American canning industry) is doing him quite a lot of good, but in his second statement he evidently overlooked the barnacles.

Barnacles head the list of those organisms that foul ships' bottoms. Acorn and stalk barnacles are both guilty, but the acorn (not the same species as the rock barnacle but very similar) is the worse. This may not seem a *very* serious crime, but it is the cause of a vast expenditure of money and labour that would otherwise be directed into more profitable channels. More serious still, it causes the loss of human lives. One must apportion the blame, of course; other animals, and various seaweeds fasten on to the hulls of ships, but of this nefarious gang barnacles are the worst.

First let us look at the loss of valuable fuel caused by barnacles, etc., while a ship is at sea. As an example, take a destroyer of about two thousand tons displacement. It leaves dry-dock clean and probably with a coat of anti-fouling paint, but careful estimates show that at the end of six months in the tropics its maximum speed will have been reduced by four knots and its fuel consumption at top speed increased by 100 per cent. (In temperate waters these figures can be halved.)

Secondly, of course, come the expenses of dry-docking and cleaning a ship's hull. I do not know what these expenses are but they must be great; dues, cost of labour, cost of maintaining a ship and its crew when idle, and cost of materials. And as regards materials, anti-foul paint is expensive in itself. It contains poisons, usually metallic poisons. These are given off slowly in the hope of preventing the settlement of colonies. They must not be given off too slowly or organisms would settle in spite of their presence, while if given off fast, they are the sooner gone. At the best an anti-fouling coat of paint has only a limited life of usefulness.

It is by slowing down the speed of ships that barnacles and their fellow-foulers may cause loss of life; ships have been driven on to rocks, been overtaken by typhoons and brought to grief in other ways by lack of the necessary margin of speed, while in times of war I need hardly give illustrations to show how vital it may often be to have sufficient speed to avoid an enemy.

As an example of the adequacy of the cement they use (to say nothing of their resistance in other respects) barnacles have been found on ships' propellers after a long voyage. Most of us have had from time to time uncomfortable sea trips, but I can imagine no more uncomfortable a journey than when travelling on the blade of a propeller!

The barnacle came from free-swimming forebears that had intelligence of a kind and led an active, hunting life. It might have risen in the world. We are inclined to regard evolution as a gradual advance, a slow mental and physical improvement, yet half the animals in the world today have turned at some point and left what brains and energy they had behind them. The barnacle is one. Yet the barnacle survives, and survives magnificently, and survival is the acid test.

CHAPTER 9

THE MOLLUSCS

OF THOSE comprehensive groups of the animal kingdom called phyla, that known as Mollusca has first place as a sea dweller. Unlike Arthropoda, *all* its classes belong to the sea. A few species, it is true, have adapted themselves to land, notably the snail and that *bête noire* of gardeners, the slug. Some have also taken to fresh water, but all the classes of this huge phylum live chiefly in the sea. The name mollusc is, as a rule, associated only with those shells one finds by the seashore, and with the edible shellfish, all stationary or at the best slow movers; it is not generally realised that this group contains also the octopus and one of the fastest and largest of sea creatures, the squid.

There are many collectors of shells, from small toddlers to erudite conchologists, and there are many shells. Anyone can find dozens by the sea within a few square yards, whilst a comprehensive book on the subject bewilders one by a seemingly never-ending list. The illustrations are no less awe-inspiring, showing shells of every size, shape and colour by the hundred. A conchologist who knows all the shells is rarer than a philatelist who knows all the stamps. Incidentally a rare shell, like a rare stamp, is—or used to be—worth a considerable amount of money. Many shells are of great beauty, and that is why little girls (and grown people) will search so assiduously for them, and for this reason they can be as interesting to the completely ignorant as to the expert, and since I belong to the former class, and also prefer the living to the dead, I will pass on to the other mollusca classes.

Octopuses, cuttlefish and squids belong to the class called Cephalopoda, a class confined entirely to the sea. The octopus has gained not a little notoriety as a manslaughterer. This reputation was probably initiated by Victor Hugo's hair-raising description in his *Toilers of the Sea* (he gave the impression that an octopus would swallow a human being like a human being swallows an oyster). Since then there has been a general idea that any diver, in a suit or not, unfortunate enough to meet an octopus will never be seen again. Those eight deadly arms, each armed with a double row of suckers, will twine around him in an embrace from which there is no escape. The octopus's beak will then be inserted and the diver will be eaten, or at any rate sucked dry. A horrid death. I do not wish to disappoint those who like horror stories, but the octopus is not really like that. It is, as a rule, only the size of a plate, or smaller, very timid (except with crabs and lobsters its own size), and very persecuted, living in fear of its life from various creatures including certain eels that seek out octopuses and eat them, after a fight in which (by various tricks) they have every advantage.

The octopus (when not hiding) walks about on the sea bed, using as a rule for pedestrianism only three of its eight legs, and it is looking for edible crustaceans, not human divers. Probably even the sight of a human diver takes years off its life. Nevertheless, if a human diver (not in a suit) were to brush against a *large* octopus he might be seized by a creature that thought itself trapped or attacked, and held until he was drowned. Indeed, divers *have* been killed in this way.

There are species of octopuses that attain a span of six feet, including of course the long arms or legs. A man who possessed one of those underwater swimming outfits told a friend of mine that he often deliberately sought out octopuses. There was no danger, he said, provided you kept the beak away from the body, and the arms with their suckers were easily pushed off.[1] He did not, however, give the dimensions of the octopuses he fought, and under water they probably appeared larger than they actually were. Some

---

[1] Other aqualung divers have *tried* to make octopuses hold on by twining a tentacle round one of their arms, but the octopuses withdrew their tentacles as soon as possible.

large specimens exist in the Pacific, but, as I say, normally the octopus is not by any means a large mollusc.

Although its normal method of progression is walking, if attacked the octopus can go backwards at speed. When doing so it emits the famous smoke-screen of black ink. This is supposed to baffle pursuers, the octopus changing direction under cover of the screen, while the pursuer charges on. And this may be the object of the screen, but there are two other theories. One is based on the re-

Octopus attacking Crab

markable colour-changing gifts of octopuses and squids. Many animals can change colour. The chameleon is a noted example. But the change is usually slow. A certain spider can change from white to pink or green if placed on these colours, but each change takes about nine days. The octopus can change colour immediately. Normally it mimics its background but it can also become colourless. And it becomes colourless after discharging its ink cloud, which has a definite shape and, for a time, looks concrete. So there are those who think that the ink cloud serves as a dummy which is attacked by the pursuer whilst the now almost invisible octopus makes its getaway. Another theory is that the discharge affects the sense of smell in a pursuer. In an aquarium a moray eel was observed to launch itself at an octopus, which dashed off and discharged its ink when the eel was a foot distant. After that the eel seemed bewildered, and even when it touched the octopus did not attack, nor seem to realise it was there. This condition lasted for an hour. The discharge of the African polecat has a rather similar effect on a pursuing dog. After receiving the nauseous fluid in its face the dog certainly cannot smell, indeed it seems hardly able to breathe.

There is no need, however, to argue as to what is the object of the ink discharge. It may give a triple protection.

Most animals have intelligence of some sort and the octopus is no exception. It has a memory. Like most of us it has been put through tests. One specimen in an aquarium, presented with plates of different colours, soon learned that its food was always served on a plate of a certain colour. Another test was to present two plates marked with diagrams (a circle and a square for instance), both containing food, but one giving a faint electric shock. The octopus would soon learn to avoid the plate giving the shock, and thereafter when the two plates were presented (neither containing food or shock) make for the diagram that used to have the food and avoid the other that used to give the shock. After this new figures were learnt, though with increasing difficulty as they kept changing. The octopus would remember the previous figures also, though not indefinitely. The most difficult test was a square marked on both plates, but put in different positions, one normal, the other with the corners top and bottom diamond-like. The majority of octopuses could learn the difference, but not all—the proportion being about four out of seven.

After a continuation of these tests several octopuses became neurotic and would not come out at all, while others, between tests, tried madly to get out of the tank. In other words, they had a nervous breakdown.

An operation on the brain—the removal of the vertical lobe—resulted in the octopus losing the power of associating electrical shocks with the figures and going for either plate indiscriminately.

An octopus will stalk its prey with all the cunning of a cat. It will also sit for hours by some bi-valve, waiting for the shell to open. It is also said that when the shell does open it will often push a pebble between to prevent its closing. Whether this is true I do not know; the main point is that an octopus can learn that sooner or later a scallop will open its shell.

Mother-love is the last thing one would expect in an octopus, yet it will brood over its eggs and keep water gently circulating round them. A female in an aquarium refused all food whilst guarding her eggs; indeed, she was annoyed when any food was given to her and always went off with it and threw it away. When the young hatched she displayed the same devotion but not much intelligence, for she continued to brood over the empty capsules. She also continued to refuse food until she died of starvation, a pitiful sacrifice for a few empty egg-shells.

But observations in an aquarium are not perhaps wholly reliable. On the face of it the octopus seems a fairly good subject; it likes a quiet life with crannies to hide in. But we still do not know how it feels about confinement and human onlookers. At any rate, captive octopuses often begin biting their arms as a worried man bites his finger-nails, and may finish up (though well supplied with food) with hardly any arms at all.

Having mentioned food served on marked plates, the eyes of the octopus (and his relatives) deserve remark. They have contributed greatly to his sinister reputation. "Baleful, glaring eyes" is the usual description, and certainly two such large eyes seem out of place in a creature of this sort. The strange thing is that they are practically the same eyes that we have ourselves. They are not faceted eyes like those of insects and lobsters, but the "camera" eye of man with expanding and contracting lenses and they must view objects exactly as ours do. And it is a remarkable thing that creatures so far removed in ancestry from us should have evolved the same eyes.

The squid can be described as two eyes and a bunch of legs at the broad end of a cone. The cone has two triangular fins on each side, and a large squid placed perpendicularly, pointed end upwards, would look like one of Hitler's V2 rockets that used to be exhibited in Trafalgar Square towards the end of the last war.

And, in a way, the squid is a rocket. It is jet-propelled by the muscular ejection of water, is shaped like a rocket, has vanes on each side like a rocket, and moves not with the speed of a rocket, but very fast. It moves so fast that it has been known to shoot out of the water altogether and land on the decks of steamers. When speeding jet-propelled it moves backwards, pointed-end first, but it can also swim in the normal way by movements of its fins.

It has ten arms against the eight of the octopus, the first pair being very long and used as a rule for capturing prey. Most species are denizens of the middle or upper waters, travelling in shoals and living on fish and crustaceans. Observers from some ship who have seen them overtaking a shoal of fish on the surface have been spellbound by their speed, ferocity and voracity.

Certain squids are illuminated, bearing lights along the mantel and arms, inside the mantel cavity, and around the large eyes. "Very beautiful" is the description given of such squids but there must be a touch of the sinister about them. We shall meet them again in the next chapter.

The general run of squids is fairly small, but gigantic specimens exist in deep water. We have no real knowledge of these creatures. No man has seen one, and but for the evidence provided by the sperm whale we should be justified in regarding them as legendary —like the sea-serpent. Certainly on one or two rare occasions a large squid has been washed ashore in a rotting condition to the surprise of all who saw it, but compared with others that lurk in the depths

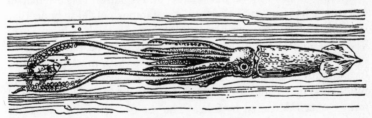

Squid capturing prey

these washed-up ones were minnows. Trawlers do not trawl at great depths, and if they did they could not possibly bring up sea monsters weighing tons, so direct data is next to impossible.

But sperm whales dive to great depths, and they dive to get squids and some of the squids they encounter down below are enormous. This is shown by the scars they receive in the fight and which they carry for life, also from remains found from time to time in their stomachs. It is easy for those with knowledge to reconstruct the whole animal from these remains and from time to time this has been done by experts. The largest squid yet found in this way measured fifty-two feet, which is a length that almost equals that of the sperm whale itself, but there may be, in fact there must be, specimens larger than this. They are unlikely to be seen, for no whale could subdue them. On the contrary, they would be the victors: their suckered arms would hold on to the whale who, like any other mammal, drowns if kept under water beyond a certain limited period. As for those toothed whales smaller than the sperm whale, squid-hunting in the deep must be a dangerous game.

A Squid v. Sperm Whale contest, if it could be filmed, would easily head the list for exciting heavyweight fights and would be much more spectacular than the substitutes artificially concocted in Hollywood.

After the squid, the cuttlefish comes as an anti-climax, closely related though it is. (The name "squid," I think, is very inappropriate: one of the fastest and largest and most voracious creatures in the sea deserved a more dignified name.) The cuttlefish is rather like a normal-sized squid, but broader and squatter. It is peculiar in carrying a hard internal skeleton, the famous cuttlefish-bone that canaries and budgerigars like to nibble and that is sold in every cage-bird shop. This light, white, oval disc can be found washed up on almost any beach, which shows that there must be millions of cuttlefish in the sea, though they are rarely seen. The bone is the inner core and before it is washed up the outside body has rotted away or been devoured by fishes. The use of cuttlefish-bone is not limited to canaries and the multitudes of "budgies" that have become so popular nowadays; it is also used in tooth powders and pastes and other things. I know a man who rubs his teeth every morning with a cuttle-bone he gets straight from the beach. His teeth certainly are white, but please do not take this as a tip, for a dentist told me he must be ruining his enamel. Cuttle-bone taken internally was once prescribed for a variety of complaints and was considered invaluable for (of all things!) leprosy. There is one thing, however, in favour of cuttlefish-bone as a medicine which by no means applies to many medicines used today—it cannot possibly do any harm. It is marvellously fine and light and contains lime and iodine and mineral salts.

Proceeding to the usual question: are the cephalopods (that is the octopuses, squids and cuttlefish) of any use? (To man, of course, being implied.) The answer is, Yes. By nourishing the whales and fish we use for oil and food they are of value, but they are of direct use as food also. It is a strange thing, especially these days when increasing populations are causing us to draw in our belts, that most of America and northern Europe have never considered eating members of the octopus and squid group, though the Eastern and Mediterranean nations consider them a great delicacy. Japan alone captures and markets about 70 million kilos a year.

It is not as if cephalopods were a kind of last-ditch stand against starvation. I often ate octopus in China and found it really good. Skate is insipid by comparison.

Although not normally eaten in America and northern Europe, squids are used as bait by the fishing industry on such a large scale that they are of considerable commercial importance for this purpose alone.

We now come to the molluscs we all know well: the limpets, winkles, whelks, mussels, scallops, clams, oysters, cockles and their kind. Nearly all of them are used as food. Even the little winkle (or periwinkle) has its fanciers. I was very partial to winkles as a boy and enjoyed digging out the tiny morsel from its shell with a bent pin, even though the rear and more succulent portion usually remained behind. I still like the taste of them, but no longer enjoy digging them out with a bent pin. They talk about the impatience of youth; myself, I think, I had far more patience as a boy than I have now. But dug out with a pin (or some similar weapon) winkles have to be; therefore it surprised me to hear recently (I cannot quote the date of these statistics) that about three thousand tons of winkles are delivered to one market (Billingsgate) every year. Think of the labour of digging out the occupants from whatever number of winkles three thousand tons represents!

All these "shellfish," as they are sometimes called, belong to two classes of molluscs, the Gastropoda (conchs) and the Lamellibranchia (clams).[1] Of the few examples given above, the first three are Gastropoda and the other Lamellibranchia. The Gastropoda is the largest of all the Mollusca classes and the only one to have contributed any members to the land. Both classes, particularly Gastropoda, are very old inhabitants of the sea and have lived there without undergoing much change for from three to four hundred million years. They are much addicted to enclosing themselves with thick, hard shells, which prompts me to make a few more remarks about this type of protection.

The desire to protect oneself is very natural, and the obvious way to do this is to wear armour. It is not the perfect answer, for armour impedes movement. It thus puts the owner at a disadvantage when getting food, when defending himself against enemies, and when trying to escape from enemies. The last two drawbacks do not matter very much perhaps, for the whole idea of armour is to make the wearer invulnerable to enemies. Unfortunately, however, the suit of armour that is invulnerable to every enemy has yet to be made.

Many forms of life that put on armour abandoned it after a period. The bony fishes, as we know, tried it once, and many of the dinosaurs went in for it in a big way, as did man himself in more recent times. Like the bony fishes, the knights of old eventually gave it up. Of the dinosaurs we cannot speak, they all died out, but there is some evidence that the very heavily armoured types were beginning to lose ground.

Yet in spite of the drawbacks and its abandonment by many species, armour has had some successes, particularly in the sea where the lesser pull of gravity makes it more practicable. On land, snails and tortoises and others benefit from it, while in the sea, crabs, lobsters and other crustaceans thrive everywhere, and a glance at the innumerable shells on any beach will show how the armour-bearing molluscs have succeeded.

Having decided to wear armour the next question is how much. In the early days of chivalry knights wore so much that they were practically unable to walk and had to be carried on horses so mas-

---

[1] The word clam is sometimes used to denote bi-valve molluscs in general; it is also used for certain species, particularly the North American hard and soft shell-fish of that name, which are eagerly sought for food.

sive that they originated the present-day "cart-horse." Then the knights wore less and lighter armour, but this made them more vulnerable. That has always been the question—how much freedom of action to sacrifice for protection. In the sea, throughout evolutionary history, we have seen compromises and changes by many species in this respect, but a long time ago certain marine molluscs decided to go the whole hog and enclosed themselves in armour so thick and heavy that movement was impossible. Amongst these was the oyster. Indeed, the oyster became the most stationary of them all. A mussel can sever its moorings, move elsewhere, and grow new ones, a scallop can turn itself over and even swim at any time. An oyster stays put. Yet the oyster, as its larvæ show us, once swam about in the sea in a fairly active manner.

It would be interesting to study the life histories of the various shellfish. There is no space to do that, so I will select only one, the one just mentioned and the best known of them all, the oyster.

A problem for the wearers of heavy armour is food. If, like the oyster, a creature is to spend all its life tucked up in bed, then food must be brought to it. The food of the oyster is brought to it by the water; minute plant life, bacteria of decaying seaweed, microscopic stuff. This food passes through the gills with the water the oyster breathes, and is intercepted, gathered into a mass, and subsequently either eaten or expelled. The oyster, it may be added, has its definite feeding times.

Free food, even if microscopic, does not float about in quantity in every part of the sea, so it is important for the oyster to choose its resting place with some care. Estuaries are generally rich in edible organisms, and it is in shallow estuaries muddy or sandy, that oysters chiefly live—or in the shallow, tidal reaches of the sea.

As I said, the oyster did not always wear armour and did not always stay put. In the seas of very distant eras it swam about with the rest, getting its food like any young herring or mackerel. I said also that we know its early history from the ways of its present larvæ. This, in brief, is what happens: after fertilisation the oyster egg turns into an embryo in a few hours. At the end of about four days it begins to swim. It swims about thus for a fortnight and then sinks slowly to the bottom, where it settles. It needs, however, some firm surface or it will be lost.

By the time it settles it has grown two small shells on either side, though it is still almost invisible. At this stage it is known as a spat, and the settling of numbers is known as a spat-fall. The baby oyster, unless interfered with by man, will now never leave its position for the rest of its life.

This is a very sketchy description of a very interesting process. I have avoided too much detail because there are two different types of oysters that reproduce their young in two quite different ways. The American Oyster (*Ostrea virginica*) together with the Portuguese Oyster (*Ostrea angulata*) is one type, and the European Oyster (*Ostrea edulis*) the other. There are other species of course but these are the commonest.

We will take the American first. Its method of reproduction is on somewhat similar lines to that of the majority of bony fishes. The female exudes her eggs (about 50 million) directly into the sea and the male releases his sperm also into the sea. The eggs float.

Even thus early the first of the tremendous hazards to which all oysters are subject appears. For the male oyster is very casual about the way he ejects his sperm, so that sperm may never reach the eggs at all, or only a few of them, and an unfertilised egg dies in a short time. Countless billions of eggs perish from not being fertilised.

And the remainder are in little better state, for they are extraordinarily sensitive to changes of temperature. Cold rain or wind will kill all the eggs and embryos floating on the sea surface. Only when they can swim and get down into less exposed conditions have they any expectation of life, and that—in common with all small unprotected things—is not much.

The young of the European oyster undergo none of these initial hazards, for until they are at the free-swimming stage (about ten days) they are tucked safely inside the shell with their mother.

The male European oyster sheds his sperm into the sea like his American relative, but the female then collects it, or what she can of it, and stores it for use. After that she lays her eggs, not into the sea but into her shell, which becomes a sort of hatching shed, and later a nursery. After hatching, the young even take their first swimming practices there.

She lays from five hundred thousand to 1.5 millions eggs, considerably less than the American oyster, but their chances of survival are more than equalised by the parental care the young receive in those vital first days.

The eggs are fertilised as soon as they emerge, and quickly develop into embryos. At this stage the embryos are white in colour, and if the shell is opened a milk-like fluid will exude. In this condition the mother oyster is called "white-sick," a term which must not be mistaken for an ailment. In three days the young are making their first swimming efforts in their private bath, and after five days they begin to colour, and are generally known as larvæ. The colouring is due to small microscopic shells beginning to form. The larvæ become light grey, then bluish-grey, then darker and darker until they are almost black, at which stage the mother oyster is known as "black-sick."

At any time now the larvæ may be dispersed to sea, for the time has come for them to leave home. When they depart no good-byes are said and there is no lingering; the mother slightly opens her shell and spits them out like a charge of pellets from a shotgun. They are now nearly a fortnight old and can swim and find their own food. In another ten days they sink to the bottom. On the sea bed they lie on their right sides, on, that is, the rounded shell. The other shell, the flat one, is a moveable, hinged lid.

It is supposed that should the spat, before settling, be carried away by tides or rough weather, they will perish. And no doubt they mostly do, but occasionally if conditions are suitable, drifting spat may settle on sea bottoms long distances away and form new oyster beds. And that is how oysters must have spread in the first place.

I do not know how long oysters can live. Twenty or thirty years have been suggested. Those whose age we are able to keep a tab on are artificially reared, and disposed of as a rule when they are only just beginning their mature lives. The European oyster is "ready for the table" in about five years. American and Portuguese oysters develop more quickly, sometimes reaching a length of two and one half inches in a year.

An American oyster is either a male or a female, and stays so all its life. Not so the European. European oysters have the urge to change their sex. So the mother oyster of this season may have been a father last season, and may be a father again next season—or the other way round. These sex changes occur quite frequently during the life of one individual. So it is strange that the European oyster mother, who is only a temporary female and will shortly be a male again, should, in a way, care for her young, while the American female oyster, who has always been a female, is as casual about reproduction as the male.

In growing, the oyster has two things to consider, its body and its shell. Its body needs nitrogenous food and its shell needs lime. The shell, in fact, is merely a solid block of limestone with a few other minerals added. These two growths have to synchronise. The shell and the body must grow together and keep pace the one with the other. Normally the shell dictates, and the shell grows fringe by fringe. The shell does not grow all the time; new fringes are only added at certain definite periods, then growth stops for the time being. Artificial or unusual conditions can result in both shell and body growing more rapidly than usual, but in these cases the shell is very thin. Certain diseases can also affect shell growth, as can a deficiency of lime in the water. This will not worry the human gourmet; he does not have to eat the shell and prefers oysters that have developed their bodies at the expense of their shells, but it is bad for the oyster; enemies are for ever on the prowl looking for thin-shelled oysters.

When we prize open an oyster we see a yellowish or creamish bag (the mantle), folded bulkily like an omelet. If any oyster-eater ever bothers to think about it, he probably looks on an oyster as a stomach with, no doubt, a few crude organs attached. But if the organs of an oyster are crude, then so are our own. The oyster is a highly developed animal. It *is* edible, very edible, but edibility is not its main function in life, it is only one of its handicaps.

Like ourselves it has a brain, a heart, and a complicated nervous system, not to mention blood circulation, liver, glands, intestines, and so on, though these do not assume the forms ours do. There are four gills and these constitute the respiratory *and* the food-collecting apparatuses. A network of small blood-vessels runs through them which join to larger veins and carry purified blood to the heart. The heart, lying in its pericordium, has two chambers, the auricle which receives the blood from the gills, and the ventricle which drives this blood through arteries to all parts of the body and then back to the gills to be repurified. The oyster, in fact, has plenty of blood but the blood is colourless. And it is just as well for our gastronomic enjoyment, I think, that its blood is not red. Were it so, I for one would not fancy such a gory dish.

The oyster possesses no eyes, though its relative, the scallop, does —about a hundred shining blue-green eyes along the mantle edge, each having a lens, retina and nerve. Some cockles, too, have eyes.

The shutting of the oyster's shell is operated by strong muscles and ligaments motivated by nerves that receive their impetus from the brain. On the inside of empty oyster shells can be seen dark-coloured spots; these are where the muscle fibres were fastened.

The oysters we generally see, on fishmongers' slabs, for instance, give the impression that the natural state of the shells is the closed position. We rarely see an oyster that is open, and if we do we very rightly view it with suspicion. But the natural position of the oyster is the open position; the powerful muscles operate only to close the shells, and to keep them closed. In its natural surroundings the oyster only closes its shell (with a snap) when danger threatens. The oysters we usually see keep their shells closed because they are alarmed by the unfamiliar conditions.

Many people have the idea that when an oyster is opened it dies. Indeed, some refer to the muscles that close the shells as the "heart," and think it *is* the heart of the oyster. And many who deal in oysters will state that severing the muscle of the shells, as with an ordinary oyster-knife, kills the oyster. An opened oyster is no more dead than is a ham-strung man.

The mantle of the oyster is a fleshy covering, delicate and sensitive. Anything abrasive would damage and hurt it. Therefore the oyster lines the inside of its shell with one of the smoothest linings found in nature—mother-of-pearl. Inside this perfect glaze the oyster can move and twist about without fear of any scratch.

Any irritant substance such as a bit of grit or sand is expelled with a violent puff, and this may happen several times a day. But occasionally, only *very* occasionally, a bit of grit lands in a place from which it cannot be expelled. So to render it innocuous to the delicate tissues, it is coated over with the same material the oyster uses to line its inside shell. This coating process goes on automatically so long as the substance is there, and in a long time (twenty to thirty years perhaps) a large-sized pearl results. If ever extracted by man this pearl will fetch a big price, and rightly so, for it is rare, beautiful, and took a long time to make. Cultured pearls take much less time to make because the centre, artificially introduced, is comparatively large itself and only needs just covering over. Real pearls and cultured pearls, therefore, may be likened to genuine silver and silver plate. But the introduction of cultured pearls knocked the bottom out of the market for real pearls.

The "pearl oyster" that supplies (or used to) most of the pearls on the market is not really an oyster but a large scallop (*Margaritifera*). It exists in quantities at diving depths in certain tropical seas. I do not suppose it makes more pearls than other oysters—pearls in oysters is a rare complaint like gall-stones in a man—but it is bigger and is not used for food and has a chance to live longer. Finding a bed of these molluscs does not mean that a man has made his fortune. It is only in boys' adventure stories that the hero collects a few pearl oysters and then, hotly pursued by rivals, departs a millionaire.

Any of the edible oysters *may* contain a pearl, though the prospects of finding one of decent size are not bright—oysters on cultivated beds are not allowed to live long enough. I have found several seed pearls, and once I got a big thrill. At a mess in Newchwang, China, we used to get oysters from time to time for lunch. They were Ningpo oysters, Ningpo being a Chinese port on the coast well known for its small but delicious oysters. One day we were each served half a dozen opened oysters. I ate a couple but when tackling the next bit into what felt like a piece of brick. I took it from my mouth and found an oval pearl the size of a large pea. It passed round the table for inspection, and its passage was watched by the goggle-eyed Chinese waiter standing behind. These, I may add, were the days when a pearl was a pearl and the cultured article was unknown.

After lunch we went to the office. Taking no risks (as I thought) I took the pearl to my bedroom and thrust it in the back of a drawer full of socks. When I came back from the office it was gone. I had shut the door and no one could have seen me hide it, so the search in my bedroom after we had left must have been intensive, yet, on the face of it nothing had been disturbed. We had the staff up for questioning, but of course they knew nothing. If only I had not been so careful and just put the pearl in my pocket. However, I expect I would have lost it one way or another before long.

Oysters are no more dangerous as food in summer than in winter. (In fact, oysters hate cold and are often killed by it.) They can never get tainted, for we eat them alive and they advertise their death by opening their shells. This is well known, and even conservative Britain allows foreign oysters (particularly Portuguese) to be imported at any time of the year. Incidentally, on a hot, sweltering, midsummer day when one does not feel like eating much, there is nothing better than a dish of large, succulent, cool Portuguese oysters.

Although the normal mortality rate in adult oysters is high, they do not appear to be subject to many diseases, nor to any fatal internal parasites. The big mortality is chiefly caused by flooded rivers that bring too much fresh water into the estuaries, or bring down silt that smothers the beds, while gales do the same thing with sand. Changes of temperature also take a heavy toll. Even shortage of food claims its victims. The "fattening beds" of the oyster farmers are not beds where oysters are planted out and fed artificially like pigs in a sty, they are natural sea reaches that happen to be rich in phosphates and nitrates, and they are not at all easy to find by the experts in the first place. And when found, there are bound to be some seasons when, for no apparent reason, they fail to give much food. There are really vintage years for oysters just as for wine. In a good vintage year oysters are fat and full of flavour.

Many people view oysters, as food, with misgiving, and are prepared to get typhoid or poisoning if they eat one. Such people should never eat oysters, auto-suggestion will not give them typhoid, but it may give them all the symptoms of poisoning. And there are those whose systems are genuinely allergic to shell-fish. Normally the oyster is one of the cleanest of creatures, but some of the estuaries where it is planted are at times contaminated by bacteria from sewage. No bacteria of this kind has any effect on the oyster itself, it is immune, but it may infect human beings who eat it. In a way, then, an oyster can be a "carrier" of typhoid, but there is this difference between it and a human carrier, the human carrier is always infectious (unless steps are taken) while the oyster, if placed in clean water, rids itself of any bacteria in a day or two. I may add that there is strict supervision in all the countries where edible oysters are specially cultivated, and on the whole an oyster is much safer food than any exposed cake in a tea-shop in summer. The Ningpo oyster I mentioned, however, had no supervision that I know of and the waters near any Chinese port are far from clean (though the Chinese, rightly, look on their sewage as gold and do not let any more of it than they can help get into the sea). Yet, of the many who ate these oysters I never heard (during a period of thirteen years) of one being ill from this cause.

Though the oyster has few diseases it has many enemies. We have mentioned some already—starfish, rays, octopuses—and have no intention of going through the whole formidable list, but we must not pass over another mollusc, the carnivorous, insatiable whelk. The whelk is the rat of the sea. Although it has a fairly heavy shell, it moves about easily, seeking what it may devour. It causes loss to fishermen by eating their fish in the nets. It kills and eats crabs and lobsters in their pots (and is itself caught by baiting a pot with a live crab). Nothing alive seems to come amiss to it, and it often works havoc on an oyster bed by gnawing into the shells of the oysters and sucking out the contents. It does the gnawing with a rasp called the radula which is a ribbon of closely-set teeth. A nasty creature, the whelk, without even the saving grace of being eatable. I am glad to say that not everyone agrees with me here and immense numbers of whelks are sold at seaside stalls and eaten by holiday-makers so that large numbers are caught and taken away from the scene of their nefarious activities, but I still maintain the whelk is not eatable—I would as soon chew a rubber tyre. Another mollusc is dangerous to oysters in a different way; that prolific breeder, and haphazard, moveable settler, the mussel, can invade an oyster bed and smother the oysters by sheer weight of numbers. Crabs also do damage by cracking open and eating young oysters and thin-shelled specimens. In short, the oyster in its heavy shell is not as safe and carefree as might be supposed. You and I experience much more difficulty in getting at its inside than do a host of other smaller creatures in the sea.

If you visit an oyster "farm" and have a talk with the supervisor and learn what has to be done before oysters can be reared to the eating stage, and hear of the many losses and difficulties, you will wonder (1) how oysters could ever have been cheap, and (2) how oysters ever managed to get along before man took them over. They *did* manage, however. Long before any mammal had evolved, in the middle of the reign of the reptiles and dinosaurs, in the Jurassic period 120 million years ago, oysters lay in quantities in the shallow seas. They were not quite like the oysters we know today; for one thing, they were much larger, and had man been present then he would no doubt have found them very succulent.

CHAPTER 10

THE DEEP SEA

Earlier on I announced my intention of dealing only with the well-known animals of the sea. Few of the creatures of the deep are well known, but to omit them would be to ignore a vast region. I have mentioned those living in shallow waters, surface waters, and middle waters, but have not yet descended to the abyss. So I speak now about the depths, and by depths I do not mean the deep seas as known to the fishing fraternity who have to fish deep to get such species as haddock, hake, and others, but real depths, four or five miles down in regions of everlasting night and great pressure. As a matter of fact, the same darkness obtains at considerably higher levels and up to about eight hundred feet, but pressure is a different matter.

We, ourselves, on shore live under a mighty atmospheric pressure of fifteen pounds to the square inch, but we have no indication of this except when we breathe. Panting after exertion, the pressure forces in our chests when we exhale, and it needs effort to inflate the chest again. Imagine, then, living at the comparatively modest depth of three thousand fathoms where the pressure is three tons fifteen hundredweights to the square inch.

When the *Titanic* struck an iceberg and sank, I read in a certain paper that this great ship, in common with all ships lost in deep waters, would sink so far and no farther. At a certain depth, the paper said, water pressure would prevent further descent, and that if it *were* possible for it to sink lower (which it was not) it would be flattened. I think this was a local paper, and the *Titanic* sank a long time ago, but the same belief is held by some today. It need hardly be said that anything heavier than water will sink to the bottom however far off the bottom may be, for it is impossible to compress water except to an infinitesimal extent, and although the pressure increases, the density of the water remains practically the same.

Nor will a ship be crushed in at any depth. A certain amount of "squeezing" will take place in the softer material. Wood, for instance, carried down and then recovered, will never float again, but sink like metal. The ship itself, however, will remain intact, for the pressure inside will always equal the pressure outside, and the two will cancel each other out. Only sealed-off, water-tight spaces will be crushed in, unless, like a bathysphere, made sufficiently strong to resist high pressure.

Similarly it was once taken for granted that no life could exist at such depths. Any animal would be squashed to pulp. It did not seem to occur to anyone in those days to wonder why we were not crushed also, for if pressure kills, then fifteen pounds per square inch will be as fatal as three tons fifteen hundredweights. In either case the subject will be pressed to death if unable to counteract the pressure outside. But, as with the ship, the outside pressure is countered by the inside pressure. If any animal were to be killed by air or water pressure then its inside would have to be a vacuum, or, at any rate, sealed off.

So at any depth or pressure, animals can live. They can even be transferred rapidly from a great to a light pressure without harm from pressure itself—though other factors such as temperature often make this change dangerous. An exception is to be found in those fishes with enclosed air bladders. This air bladder is so regulated that when pressure decreases, the surplus oxygen can be absorbed by the blood, but if the change is too quick, as, for instance, when a fish is hauled up from deep water to the surface, the bladder swells as the pressure decreases and may shoot out of the mouth like a monstrous balloon or explode inside like a bomb, blowing the fish into fragments.

Apart from the problem of pressure for life in deep water (which, as we have seen, is really no problem at all) another difficulty might seem to exist for the abyss dwellers: how do they get their oxygen?

The only source of oxygen in the sea is from the air at the surface and from plants at the surface not more than three hundred feet down. How can this oxygen get down to twenty thousand feet and more? Surprisingly, there is almost as much oxygen at these remote depths as there is in the upper surface; indeed, there is more than at certain levels far above.

As we have said, all the oxygen in the sea comes from the surface layer down to about three hundred feet. In this layer the microscopic plants are busy at their work of taking in CO_2 and giving out oxygen by photosynthesis, while at the surface itself atmospheric oxygen is taken into solution direct. The result is that this three-hundred-foot surface layer contains very nearly as much oxygen as the water can hold. What remains, then, is merely a matter of distribution. This is effected by currents and rises and falls, and the action of winds and waves. In effect, the sea is being continually slowly stirred like soup in a saucepan. Cold conditions on the surface send the surface layer sinking to the bottom; there are corresponding upwellings of water from below; currents bring water from other parts of the sea; the oxygen-charged cold water from the poles creeps along the sea bottom almost to the Equator. The sea, in fact, is a vast water-aerating (and nutrient-making) machine. Only in certain areas, as in some Norwegian fjords, cut off from sea movement by barriers at the entrances, is the bottom water deficient in oxygen, while the Black Sea, separated from the Mediterranean by the shallow and somewhat stagnant Bosphorus, has no oxygen at all at its lowest levels and consequently no life there to consume the fallen animal and vegetable remains that fall to the bottom and lie in poisonous decay (the source perhaps, later on, of some rich oilfield).

So neither oxygen shortage nor pressure worry the bottom dwellers of the ocean, but there remains a third difficulty. This (as usual) is food. On the face of it, it would seem impossible for any of them to get any food at all. The only food they can obtain must come from far above in the shape of dead organisms or fragments of decayed meat. But consider the difficulties: a saw fish, say, slashes into a shoal of fish and leaves countless bits and pieces which it cannot retrieve. These sink. But on their slow journey downwards they have to pass through tiers and tiers of hungry fishes living at various depths, and after even only a few hundred feet it is unlikely that any will remain. But suppose some of these bits and pieces do run the gauntlet and continue their journey. They have some miles to go yet and soon the action of bacteria comes into play; the scraps rot and corrupt into liquidity. It is the same with the dead microscopic organisms, vegetable and animal, that are for ever falling like rain from above; these, too, are attacked by bacteria and if they arrive at the bottom at all and are not completely liquified, they arrive only as calcareous skeletons, of little value, one would think, as food. Many of the strata, of course, are made up of these accumulated microscopic skeletons, but even limy skeletons cannot survive the journey to *great* depths.

In view of these difficulties it is often stated that the inhabitants of the deep live chiefly on each other. But it cannot be explained away like that. A large number of them, and all the larger forms, *do* live, not chiefly but entirely on each other, but there has to be a fundamental source of food, and there have to be creatures whose thankless task it is to take in this food and distribute it amongst the others in their own persons. And, whether they "live on each other" or not there has to be enough basic food to support the whole population that live at the bottom.

So, in spite of all the difficulties, food of some sort obviously does filter down from above to the remote depths and settle in the ooze. It may be vegetable, or animal or both but it gets to the bottom, far from fresh, of course, but eatable.

There was a time when no one believed that life of any sort could exist in the sea at a greater depth than about two thousand feet. It was taken for granted that the deep ocean bottoms were muddy stretches as barren as in the days before life came to earth. Then came the laying of submarine cables by ships like the *Great Eastern*, and these cables often broke, or on account of some defect had to be fished up again. Those recovered from depths showed evidence of marine life, and this roused the interest of an oceanographer, Edward Forbes. It was largely due to him that in 1868 a naval vessel, H.M.S. *Lightning*, and the next year another vessel, H.M.S. *Porcupine*, were fitted out with the best apparatus then available for deep-sea dredging and sent off to dredge in deep waters.

At that time it was thought that the bottom of the deep sea was a vast plain of ooze where all inequalities had been flattened by pressure, and hills levelled as worm casts on a lawn are levelled by a roller. Ooze the bottom undoubtedly is[1]—most of it—and there are vast plains, but it is not universally flat. At one time a single sounding at only moderate depth took a whole day. Very deep soundings, thanks to the echo device, can now be made in a few minutes, and these show that the bottom of the sea is shaped much like the surface of the land: plains, mountain ranges, hillocks, ravines and the rest.

The results obtained by the ships *Lightning* and *Porcupine* were not spectacular, but they showed that animal life (though not, of course, vegetable) was fairly abundant in those supposedly lifeless regions.

The scientific world was agog. What *sort* of life existed in the depths? Surely many of the creatures there must be the left-overs, the armoured fishes and other ancient forms now only seen in rare fossil pieces, the primitives of the early eras, the remote begetters of present life. So, in 1872, H.M.S. *Challenger* was fitted out and set off on her famous three-and-a-half-years' voyage to probe thoroughly into these matters. The expedition verified beyond dispute the existence of life at great depths, but, alas, found that this life was comparatively modern; the bottom of the sea contained no living fossils. Just as there are no dinosaurs left in any forgotten corner of dry land, so there are no Cambrian relics swimming in the dark unfathomed depths of ocean. There is no hiding away from the march of time—at least, not for that long.

Nevertheless, if the results were disappointing to some, they were sufficiently startling. A collection of creatures was brought to light, then and subsequently, that might have featured in the worst nightmares of a wood-alcohol addict; creatures with huge mouths and long needles for teeth, illuminated eyes on stalks sending beams of light through the darkness, illuminated mouths and insides, and

---

[1] The chief oozes are calcareous and siliceous remains, and red clay (which forms the bed of the deepest depths). These have been laid down extremely slowly—on an average at the rate of 1 cm. per thousand years.

bellies incredibly distended with a living victim inside three times
the size of the small fish that had engulfed it and plainly visible
through the skin. These and other monstrosities.

Unusual, very unusual, though many of them are, the present
deep dwellers probably only invaded their habitat in Cretaceous
times, a mere 70 million or so years ago. Before that time the ocean
beds may have been unoccupied, or more probably were occupied
with remoter forms that died out. The present inhabitants came
from above like so many Satans falling from Heaven. Just as the
fishes that first invaded land were rejects, so are most of the deep-
sea dwellers, for no fish or any other form of life would have left
the comfortable well-fed existence in the upper stories for a miserable existence in the basement unless it had to. In their new horrible environment the fishes changed, as outcasts often do, and
became unrecognisable from their parent stock. They assumed new
shapes and became smaller. Food was short so those that "lived on
each other" could not afford to miss any opportunity and grew
enormous mouths and wicked teeth. There can be few long-drawn-
out fights in the sea underworld, so murderously armed are the
majority that a meeting between two of them must be as quickly
over one way or the other as a meeting between two gangsters
using revolvers. Large mouths down there also are almost as necessary as long teeth. Food, I repeat, is short, and the victor must be
able to bolt down the victim, and as quickly as possible—some
other candidate for a meal or death may come along at any time.[1]

Owing to the strict rationing of food below, however, life cannot
be very congested. There are no great shoals of fish to be preyed
on by destroying shoals of other fish. Life in this underworld must
be rather like the night life in the slums of certain cities in the
old days: single assassins carrying knives prowling in the dark. Similarly the assassins in the sea basement prowl around, but for the
most part they must prowl to no effect and grow hungry. So when
they do find a victim it may have to last them a long time. Distensible stomachs are an answer to this, and few forms of life possess
a stomach more distensible than the deep-sea dweller *Chiasmodon*

[1] Probably the long teeth also act as a barrier, preventing the escape of a
swallowed victim as iron bars prevent the escape of a prisoner in a cage.

Deep Sea Fishes

niger. This is the fish I have already mentioned that can swallow a prey three times its size and keep it in its stomach. It is as if a wolf were to swallow whole a large sheep.

The deep-sea creatures, strange enough in appearance, most of them, as they are, are also able to light themselves up. How this lighting-up is accomplished we do not really know. So to talk about the bottom sea being a region of everlasting darkness is not quite correct; illuminated fishes, etc., are always moving here and there. In fact, if a number of these were to gather together that place would resemble Broadway or Piccadilly Circus at night, a sort of fairyland, for many of the lights carried by the bottom dwellers are coloured. The pattern of lighting varies. Some species have a row of lights along their bodies, others whole tiers of lights along their sides making them look like ocean liners at night, and which they can switch on or off as they desire, some have illuminated circles round their eyes and mouths, some illuminated heads and faces, some are illuminated all over, some glow from inside. Particularly strange are the fishes carrying a light at the end of a long rod in front of them. Presumably other fishes mistake this light for some small organism and dash at it, only to be seized themselves by the form behind, invisible in the darkness. Most remarkable of all is Lasiognathus, the fish angler. This fish carries a fishing-rod armed with hooks at the end, and a light. Another fish seeing the light makes for it and is hooked. But a fish hooked at the end of a long rod in front would be of no more use to Lasiognathus than a carrot at the end of a stick is to a donkey—so the rod is provided with a hinge in the middle.

Many of the non-swimmers have enormously elongated legs and other devices to prevent them sticking into and being smothered by the ooze.

Present amongst the company are two old friends of ours, the squid and the hermit crab. Both these animals are as a rule well able to look after themselves, and they have kept up their reputation in the deep. Taking the squid first, it will be remembered that it ejects an ink cloud to escape enemies. But an ink cloud in the general blackness would be wasted effort, so certain species of squids in the deep eject a *luminous* cloud, while the hermit crab of the kind that carries two sea anemones goes one better by carrying two illuminated anemones, which not only give it protection but serve as torches too. (Other familiar forms are also present, including prawns, and Houot and Willm, in a bathyscaphe dive off Dakar in 1954, saw a six-foot shark at thirteen thousand feet.)

There is some doubt as to what is the object of this lighting-up. In the case of the angler fishes the lights act as a lure, and possibly as a torch also. What of the others? It may be just to give them light, and illumination down there is badly needed. Indeed, certain forms are equipped with headlights like a car. These lights are placed just in front of curved, glistening reflectors near the eyes and are projected as two beams of light.

On the other hand, in that lurking place of assassins, lights must often attract undesirable attention.

It may be that the lights in many cases serve as a recognition mark between the sexes, for the pattern of lighting is fixed in the various species. Probably it is difficult in those places of danger and darkness for the sexes to get together at the right time, and the light patterns may help.

In some cases evolution, as if it recognised these hazards, has brought about a very remarkable method for ensuring the fertilisation of the female's eggs. A certain species, of the angler-fish tribe, judging by all the specimens secured, possessed only females. No males were ever found, even when such ought to have been hanging around. It was rather a mystery. Where was the husband of this female who regularly produced her fully fertilised eggs? The truth, when it came to light, was indeed surprising. The male, when very young and very small, bites the skin of the female, and holds on. He holds on so tenaciously that in time he becomes embedded in her flesh and her skin grows over the aperture and imprisons him. He is now a part of her and gradually becomes even more so. Her veins merge into his. His heart and digestive system decay away (he has no need for them for her blood flows through him and nourishes him). In his almost empty body his testes enlarge and develop like some cancerous growth until his whole inside is crammed. To all intents and purposes he is now merely a bag of sperm, and the female can be said to contain both ovaries and testicles. For she can draw on his sperm as she desires whenever she lays her eggs. Never in nature has female possessiveness gone further.

We must picture the underworld of the sea as a place of intense darkness marked by moving lights. On a black winter night recently I waited for the congregation to come out of a country church that stood alone about half a mile from the village. The congregation emerged, the church doors shut, and the people went their several ways home, carrying torches. I watched the lights bobbing and weaving and gradually disappearing and it occurred to me that some such sight was being reproduced in the depths of the sea: lights bobbing and weaving and coming and going. How many of these moving lights there would normally be in a given area we cannot guess. Bathyscaphe observations indicate there are more than was once supposed. It is certainly not a blaze of light down there, but enough light is provided by the inhabitants to have enabled most of them to keep their eyes.

For most of the deep dwellers have large and sensitive eyes, and if these eyes over the years had had nothing to work on they would have disappeared. Dwellers in caves, where darkness is complete, invariably lose their eyes over a period. An eye which never sees any light *must* go.

No specimens have been brought up from the deepest places. This is not because there are none there, but because it is impossible at present to dredge at more than ten thousand to fifteen thousand feet. Figures are dull things, and when they get into the thousands convey little unless a mental comparison is made, an effort which becomes greater as the figures increase. Here, then, is a comparison which may make it easier to realise the depth of the sea in certain places: if Mount Everest (29,002 feet) were removed and dumped into these depths its summit would be a mile beneath the surface of the sea. Quite two-thirds of the sea floor lies at a greater depth than ten thousand feet, a fact that might be considered by those who complain that the earth has now no places left unexplored.

EPILOGUE

I SET out in the latter part of this book to glance at a few of the better-known of the bewildering multitude of creatures that fill the sea. There is no stopping-place for a subject so vast. But there is a limit to the length of a book of this kind, and I have reached that limit now. I know that I have left out many interesting subjects and many interesting groups of animals, but this was inevitable.

Going back to the first part of the book, to evolution, it seems a pity that, though we can reconstruct the past forms of life, we cannot even imagine those of the future. In 10 million years, or 500 million, if it comes to that—the earth is young and has a long span of life ahead of it—surprising things may come about. A new race, dwarfing man in intelligence, may shoot like a star from the ranks of the lower animals; in changed conditions the dinosaurs may regain their lost kingdom. It is exceedingly unlikely, but there are still reptiles left over to start the ball rolling; birds may oust mammals; squids supplant fish. And the insects may have a big part to play—they are already asserting themselves in no small way. But all this is idle fancy. The only certain thing is that, over the eras, the present species will die out or change.

Equally certain is the continuance of terrestial changes. The inconstant land will come and go, sinking here, reappearing there, itself contributing to changing climates and species.

But the sea will remain. The sea is the source and home of life. Like a pregnant mother she nourishes the life within her with nutrients and oxygen. And she herself appears to be alive. The thud of the surf and the roll of shingle on the beach, as measured as the ticking of a grandfather clock, is like the beating of a pulse. Even when the sea is quiet and the surface like a mirror there is a perceptible rise and fall like the breathing of something asleep.

And were the sea to go, all life on earth would go.

TIME PERIODS AND FOSSILS

(1)

THE various time periods have been frequently mentioned in the text, so I give below for reference diagrams to illustrate their sequence. These time-scales are based on the rock strata that have been laid down over the earth's history. At one time the age of these strata was computed from various data and estimates, such as the thickness, number of interior layers, etc. The result was a computation that is now known to have been very far from accurate. Then uranium came to the geologist's assistance in the shape of radio-active rocks. These are certain igneous rocks which disintegrate into helium and lead at a fixed and constant rate in all conditions, so that by assessing the amount of uranium-lead it is possible to determine the age of the rock. The stratified rocks can then be correlated with the dated igneous rocks, which serve as a kind of check. This is to put the matter, for the sake of brevity, in an oversimple way: there are other factors and many complications. Those who are interested should study some book on the subject, such as Zeuner's *Dating the Past*.

The whole of time is usually divided into four *Eras* (sometimes called *Epochs*). They are represented below in diagram.

2500	500	175	70
PRE-CAMBRIAN	PALEOZOIC	MESOZOIC	CENOZOIC
A	B	C	D	E

"A" marks the beginning of the world, and the figure above the estimated number of years (in millions) from the present time— B.C. in other words. It need hardly be said that these figures are very, *very* approximate. (At one time these divisions were known as Primary, Secondary, and Tertiary. A to C was Primary, C to D Secondary, and D to E Tertiary, and these names have never entirely been dropped.) The Pre-Cambrian era is often divided into the Archeozoic (2500 to 850 million B.C.) and the Proterozoic (850 to 500 million B.C.).

The three eras that follow the Pre-Cambrian are divided into *Periods*. Taking them in order, the first, the *Paleozoic* (B-C) era has the following periods:

500	450	375	340	275	200	175
CAMBRIAN	ORDOVICIAN	SILURIAN	DEVONIAN	CARBONIFEROUS		PERMIAN

B   C

In America the Carboniferous period is subdivided into Mississippian and Pennsylvanian.

The *Mesozoic* (C-D) era has only three periods:

175	150	100	70
TRIASSIC	JURASSIC	CRETACEOUS	

C   D

The Cenozoic (D-E) era has six periods:

70	30	20	10	1 million	10-15,000 years
EOCENE	OLIGOCENE	MIOCENE	PLIOCENE	PLEISTOCENE	RECENT

D   E

The Recent, or Holocene, period, the period in which we live, has its conventional beginning at the time of the decline of the last glaciation. Radio-activity has again served as a checker-up of dates but not this time in the form of uranium. It has been found that a very small but fixed proportion of the carbon dioxide in the atmosphere is radio-active (having been formed, it is believed, by cosmic rays acting on nitrogen molecules). All plants absorb CO_2 together with its radio-active proportion. The radio-active proportion is known as Carbon 14 (14 being the atomic weight: ordinary carbon has an atomic weight of 12). And since all animals, directly or indirectly, live on plants, their tissues, too, contain C_{14}. It was Dr. Libby of the University of Chicago who showed in 1950 that C_{14} would click in a sensitive Geiger-counter and establish the age (within a very small margin of error) of any plant or animal remains. But only up to the age of 25,000 years, for after 25,000 years C_{14} disintegrates entirely. When the glaciers in our last ice age reached the southern limit of their advance they pushed over and engulfed many forests. The fossilised trunks in some cases still remain and tests with C_{14} show their age to be about 11,000 years, and not 25,000 as had been supposed. So the date of the beginning of the Recent period can now be put down at 10–15,000 years.

I need hardly add that C_{14} is also of great value to the archæologist in establishing the age of prehistoric remains.

Fluorine is another recent check based on the fact that all remains absorb fluorine from the ground slowly but persistently. It is, however, of comparative value only, for the fluorine content differs in different localities.

(2)

As everyone knows, we owe practically all our knowledge of past forms of life to fossils and the subject is a vast one. So I do not propose to go into details about various "finds" nor into technicalities about digging for fossils; I only propose to make a few general remarks.

What is a fossil? The word means something that has been buried, so that if we take the meaning literally we ourselves and our dogs and cats become fossils immediately after internment. To get over this ambiguity Sir Albert Seward ruled that burial by man was a disqualification; only burial by nature counted. This puts in some doubt the status of a bone buried by a dog or a dead insect buried by a beetle, but that is only making childish objections.

Many class as fossils only matter that has been turned into stone after long natural processes, but since other natural processes are almost as good preservers of dead organisms as petrification this distinction is too limited.

The word "sub-fossil" has been brought into use and applies to remains dating no farther back than the end of the Pleistocene period, but this does not help very much.

In short, there is no strict ruling and every man must be his own interpreter of what constitutes a fossil.

Even burial is not a pre-requisite. In arid deserts animals have been preserved by desiccation, remaining above ground for very long periods. Subsequent burial by sand storms would preserve them longer but they are still fossils.

Great age is not necessary. The great auk was exterminated about the middle of the last century. Preserved eggs and skins of this extinct bird are occasionally found and are classed, by most, as fossils. The dodo was exterminated in the seventeenth century and specimens are still found preserved in bogs. Bones of the giant moa of New Zealand (extinct about six hundred years) are found occasionally in swamps and caves. These remains are usually referred to as fossils.

Nature has many preservatives. One is peat. Four hundred years ago, in Scotland, a traveller became lost in boggy country. He was not found until just recently. He had been trapped in a bog and entombed in peat. He was almost as "fresh" as he was the day, or more probably night, when he sank into the mire. The peat had preserved everything including his tweed clothes which, it is said, almost looked as if they only needed a visit to the cleaners to make them fit to wear again.

Another long-dead human being was recently found in peat. He had been dead four thousand years and his fine, strong features were almost as they had been in life. He had been strangled by a rope and was judged to be a chief who had undergone some ceremonial sacrifice.

Another of Nature's preservatives is to be found in any modern

kitchen in the shape of a refrigerator. Cold plus burial (the latter
to protect the subject from desiccation, temporary spells of warmth,
and the attentions of carnivores) is one of nature's most successful
preserving processes. A well-known example is the discovery by
hunters in 1900 of a massive woolly mammoth in eastern Siberia.
The monster had been buried in frozen soil and the preservation
was complete down to the uncouth coat with its long black hairs and
inner brown wool. Even the flesh was not only intact but eatable.
Wolves had devoured part of the trunk and the hunters used part
of the rest as food for themselves. Not only this, the corpse was in
such a condition that an autopsy was possible. Some limbs were
broken, there was clotted blood in the chest and uneaten grass between the jaws, all showing that death had come unexpectedly—
probably from a fall into a crevice. So any jury could conscientiously
have recorded a verdict of death by misadventure.

Yet this animal met its fate some twenty thousand years ago.
Quite a number of other frozen woolly mammoths have been
found in Siberia.

A preservative also found in the kitchen is oil. Oil will keep olives,
sardines, etc., fresh for a long time, especially if the jar is sealed.
So while man preserves sardines in oil, nature does the same thing
with woolly rhinoceroses. That long-extinct prehistoric animal is occasionally found in the oil-permeated soil of Galicia in Poland together with bits of its flesh and hide.

The asphalt pools of California and Trinidad must have witnessed many grim scenes. Here have been found in large numbers
the skeletons of sabre-toothed tigers, wolves, horses, sloths, elephants and others. The apparently solid surface lured animals on
to the tar, to be held in a grip from which there was no escape and
where the heaviest and strongest had even less chance than the
others. The bellowings of the trapped ones evidently drew the carnivores from miles around in expectation of an easy meal (for the
remains of carnivores in the asphalt greatly outnumber those of
herbivorous animals) only to be trapped themselves. These asphalt
pools have provided us with some of our most valuable fossils.

Prolonged dust gales have overcome many animals (often found
in numbers all facing the same way, presumably away from the
wind), buried them in sand and fossilised them.

The best preservative of all is gum, though unfortunately it can only operate on small creatures. Nevertheless, since insects are rarely preserved at all by normal natural methods, gum has been a great friend to the entomologist. Long ago on the shores of the Baltic grew forests of pine of the species *Pinus succinifera*, and these pines, as most pines do, occasionally exuded drops of gum from the trunk and branches and occasionally (*very* occasionally) a drop might chance to fall on an ant, a fly, a bee, a spider, or some similar creature going about its lawful occasions. This held the object for a time, and before it could get free another drop came, and then another. And so on, until the unfortunate insect was encased in gum. The gum then gradually hardened, and occasionally bits got detached and got into the sea, where over a long period they became fossilised into what we call amber. Much amber has been found on the shores of the Baltic and some has drifted to the east coast of England and can be found there even today, especially in Norfolk on the sands when the tide goes out, though the chances of getting a piece containing a fly are remote.

Now Baltic amber is of the Oligocene period and about 50 million years old, and the insects preserved in it are of course of the same age. Their preservation is complete down to the minutest detail and we can see as in a glass window a spider in the act of catching a fly and other dramas, just as they happened, and with the real actors and actresses, 50 million years ago.

A petrified footprint is, of course, just as much a fossil as the petrified remains of the animal itself. In a canyon in Arizona, very clearly marked, are the footprints of a large dinosaur, giving, incidentally, an awe-inspiring impression of the animals' size and length of stride. Indeed, such petrified footprints have sometimes led antiquarians to the fossilised body of the animal itself, just as the spoor may lead hunters to the living quarry.

Very ancient is the craft of making death masks. There is an excellent one of Edward I and other long-dead rulers. Nature also makes her death masks, sometimes with exquisite care. In 79 A.D. the ancient city of Pompeii was destroyed by an eruption of Vesuvius. Death came to most of the inhabitants in the shape of showers of volcanic ash that smothered and buried them. The bodies decayed and the ash hardened and gradually an empty hollow was made where the bodies had lain, so that nearly two thousand years later it was only necessary to fill in these moulds with plaster to obtain the perfect cast of a citizen just as he lay after death had overtaken him. There is one such cast in the Geology Department collection of Imperial College. It shows a man lying sideways with his head resting on his right arm. His toes, limbs, body, face and even his hair are shown in every detail. His expression, strangely enough, is peaceful, and the man looks as if he were asleep. Dogs and other animals of Pompeii have been similarly preserved in, as it were, the negative.

When coal was first brought to the surface in quantity it changed our whole way of life. The fossils of Carboniferous-period vegetation saved and gave to us the heat and power of the sun that shone on the world 250 million years ago.

The whole of the processes of converting the luxurious forests that flourished in that long-distant warm and steamy climate are not known, but the broad outlines are all we need here. The trees (probably growing in marshes and lagoons) die and fall and are attacked by fungi, bacteria and other bio-chemical agencies and converted into a kind of peat. Mud covers them, more trees fall forming another layer, and so on. Pressure steadily increases on the lower layers and in addition a chemical process comes into play which results in the gradual elimination (the whole thing is very gradual; there are no short cuts in nature's coal-manufacturing processes)—of oxygen and hydrogen in the form of water, carbon dioxide, and marsh gas. When the pressure and chemical actions are complete, then we have coal. The product is simply trees fossilised into carbon and minerals under intense pressure.

Certain conditions, of course, are necessary for this process. Trees can be fossilised by other of nature's processes—not into coal, but stone (and the difference, in the opinion of some purchasers of coal recently, is not always very marked).

Remains of ashes show that the Romans found and used coal during their occupation of Britain, but the boom began in the eighteenth century and caused the industrial revolution, a great increase in population, and the rise and fall of nations. When one considers the amount of coal we use every day in the countless mills, foundries, trains, steamships, homes, etc., of the world, we get an idea of the extent and luxuriance of those steamy Carboniferous forests.

Can a liquid be a fossil? Definitely. And the oil and petroleum that are being used increasingly in place of coal are no less fossils than coal itself. For these oils are derived probably from the accumulated bodies of small fishes that once flourished in great inland seas that became silted over and in the course of time sank to great depths. And if these types of oil were vegetable (diatoms, etc.) and not animal in their origin it makes no difference, they are still fossils.

But in spite of the exceptions, the bulk of the fossils which have supplied us with our history of the past *are* of stone, and many and varied are the specimens that have been unearthed. And yet the chances of any animal becoming a fossil are remote. Of land dwellers modern man ought to have the best expectation in that he is generally buried straight away after death, but though you and I may have marble statues erected to us, we cannot really hope at all for the immortality that fossilisation brings. I do not know what the life of a marble statue is, but geologically it cannot be long. We have many fossils 500 million years old and, conditions being right, there is no reason why they should not last through a whole succession of such immense periods. Yet once made, and well and truly turned into stone, a fossil is not necessarily on the road to permanence. Should percolating water find it, for instance, it is doomed, and the strains and stresses of the heaving earth may at any time grind it into dust. And this has been the fate of more fossils than any that remain on earth now. For every fossil we unearth, millions lie embedded in rocks miles below the surface, completely inaccessible and, even if they were accessible, beyond anyone's powers of chiselling out of the stone.

The first step (after dying) to become a stone fossil is to get buried, and after that to get sealed up. The third requisite (in most cases) is to possess some hard parts such as a skeleton or a shell. To get buried naturally and then sealed off in a suitable medium in a short time can happen only to the favoured few, and that is partly why a fossil is such an exception.

The sea presents the best conditions for this process. It might not seem so—dead fishes, for instance, floating about with hundreds of crabs and other hungry marine carnivore ever on the lookout for a corpse to eat—but you must consider the shifting nature of sands and the silt laid down in the sea by rivers. It may be a miracle for a fish to get decently buried and coated over, but all fossils are miracles. Anyway, the survivors (survivors that is, from the point of view of fossilisation) sink to the sea bed and are given a protective upper layer of sand. Clay, or other silt may then accumulate over the sand and the strata sink, gradually solidifying under pressure into rock, together with its dead passengers. Or the sand itself may solidify, its grains being cemented together, usually by calcium carbonate.

On land, roughly speaking, the same process happens, but more rarely. The subject that escapes bacteria, flies, carnivores, and other perils to a corpse, gets swallowed in a lake or lagoon or bog, or is overtaken by a sandstorm and is buried and in due course is sealed up in some air-proof stratum that will eventually become rock or sandstone—for we are speaking now only of petrified specimens.

As for the process itself, the "turning into stone," it differs. Usually the harder parts of the organism, the bones, skull, etc., are gradually, infinitely gradually, replaced by minerals such as silica, calcium carbonate, iron pyrites. The substitution is molecule by molecule so that the replacement is complete down to the minutest detail and the subject can sometimes be studied under a microscope almost as if it were a fresh specimen on the dissecting table. The soft parts generally disappear but occasionally under certain circumstances, are also replaced in minute detail by minerals. More often the soft parts (or the complete creature) leave a mere outline on the rock.

The remains of plants, as any amateur who has done any searching for fossils knows, are more frequently found than those of animals. This is only natural; plants are always shedding bits of themselves around and make no effort to escape the traps of bogs, lagoon and other burial places which the animals are at pains to avoid.

Whole tree trunks may be petrified by being carried by some river and buried in a silting delta at the mouth. Erosion may later expose the silicified wood. In the U.S.A., in the Yellowstone National Park there are petrified tree trunks still standing *in situ*. What happened was that the trees were, long ago (in the Miocene period), buried in volcanic ash—like the citizens of Pompeii in a later period. Afterwards they were uncovered by erosion exactly as they were— except that their fibres were now stone, not cellulose.

Fossils give us much, indeed all, of our knowledge of the changing forms of life in the sea and on the land ever since life became large enough and hard enough to have, so to speak, its photograph taken in the form of a rock impression. They do more, they tell us amongst other things of climates and temperatures in distant times. If an animal or plant, for instance, is known to inhabit only tropical regions, the fossils of that animal or plant if found in temperate places will indicate that those places were, at the time the fossil was laid down, tropical, even if they are now under a glacier.

Fossils are a record written in stone, a kind of hieroglyphics that, like the Rosetta Stone, have been gradually and painstakingly deciphered, and been used as a key to our knowledge of life in the past.

BIBLIOGRAPHY

Apart from journals, the following were amongst the works consulted:

Calman, W. T. (1911). *The Life of Crustacea*. London. Methuen.

Chapin, H. and Walton-Smith, F. G. (1953). *The Ocean River*. London. Gollancz.

Dakin, W. J. (1953). *Australian Sea Shores*. London. Angus and Robertson.

De Latil, Pierre. (1954). *The Underwater Naturalist*. London. Jarrolds.

Green, N. (1918). *Fisheries of the North Sea*. London. Methuen.

Hardy, Alister C. (1956). *The Open Sea, its Natural History, Part I: The World of Plankton*. London. Collins, the New Naturalist.

Huxley, Julian. (1953). *Evolution in Action*. London. Chatto and Windus.

Jenkins, J. T. (1927). *The Herring and the Herring Fisheries*. London. P. S. King & Son Ltd.

Kenyon, Ley. (1956). *Collins Pocket Guide to the Undersea World*. London. Collins.

Marshall, N. B. (1954). *Aspects of Deep Sea Biology*. London. Hutchinson.

Matthews, L. H. (1952). *Sea-Elephant*. London. MacGibbon & Kee.

Moore, Ruth. (1954). *Man, Time, and Fossils*. London. Cape.

Morgan, Robert. (1956). *World Sea Fisheries*. London. Methuen.

Norman, J. R. and Fraser, F. C. (1937). *Giant Fishes, Whales, and Dolphins*. London. Putnam.

Ommanney, F. D. (1949). *The Ocean*. Oxford University Press.

Simpson, G. G. (1953). *The Major Features of Evolution*. New York. Columbia University Press.

Smith, B. W. (1939). *The World under the Sea*. London. Hutchinson.

Zeuner, F. E. (1946). *Dating the Past*. London. Methuen.